WILD
SCIENCE

Unexpected Encounters
When Working in Nature

EDITOR

HELEN P. WAUDBY

CSIRO
PUBLISHING

A catalogue record for this book is available from the National Library of Australia

ISBN: 9781486317639 (pbk)
ISBN: 9781486317646 (epdf)
ISBN: 9781486317653 (epub)

How to cite:
Waudby HP (Ed.) (2024) *Wild Science: Unexpected Encounters When Working in Nature*. CSIRO Publishing, Melbourne.

Published by:

CSIRO Publishing
36 Gardiner Road, Clayton VIC 3168
Private Bag 10, Clayton South VIC 3169
Australia

Telephone: +61 3 9545 8400
Email: publishing.sales@csiro.au
Website: www.publish.csiro.au
Sign up to our email alerts: publish.csiro.au/earlyalert

Cover: A silhouette of a reindeer grazing on a snowy slope (photo by BriefcaseToBackback/Shutterstock.com)

Edited by Adrienne de Kretser, Righting Writing
Cover design by Cath Pirret
Typeset by Envisage Information Technology
Printed in Australia by McPherson's Printing Group

CSIRO acknowledges the Traditional Owners of the lands that we live and work on across Australia and pays its respect to Elders past and present. CSIRO recognises that Aboriginal and Torres Strait Islander peoples have made and will continue to make extraordinary contributions to all aspects of Australian life including culture, economy and science. CSIRO is committed to reconciliation and demonstrating respect for Indigenous knowledge and science. The use of Western science in this publication should not be interpreted as diminishing the knowledge of plants, animals and environment from Indigenous ecological knowledge systems.

The paper this book is printed on is in accordance with the standards of the Forest Stewardship Council® and other controlled material. The FSC® promotes environmentally responsible, socially beneficial and economically viable management of the world's forests.

MIX
Paper | Supporting responsible forestry
FSC
www.fsc.org FSC® C001695

Aug24_01

FOREWORD

My life holds few distinctions, but one thing I can lay claim to is joyfully following some of Australia's most qualified people into bushes. Sometimes before we've even left the airport car park.

This sounds more dodgy than it is. Having worked as the ABC's nature reporter for many years, one of the things I've learnt is that ecologists may not be good tourists, but they're easily the best tour guides. It's impossible for them to not see things – in the bushes, under bushes, behind bushes.

Even marine biologists. The instant they spot something in an overlooked clump beside a boat ramp, you can bet your wetsuit that everyone will need to take a closer look.

Despite how fortunate I am to spend time with them, I'm not a scientist. Yet somehow, in the midst of these devoted ecologists, I've always felt welcomed. I suspect it's my willingness to gaze at rocks or leaves or insects – the genuine enthusiasm to take a closer smell of things before asking 'What IS that thing?'.

And while I can't help but be grateful for things I've seen and smelt and touched and sometimes tasted, field work is about so much more than peeking under bushes. Yes, it's true that when the weather is just right and the environment is revealing itself, it can sometimes feel like intellectually elevated nature bathing. But it's truly vital.

Helen Waudby writes in the preface to this book that 'field work is the foundation of the ecological sciences'. It's true even in a broader sense; without lived experience of our own, we are no more than a bunch of computer models, museum specimens or historical texts.

Field work is the tangible human experience *within* the science. It's a love poem compared to an instruction manual – what does a newly fed penguin chick feel like in the hand? What does the antechinus smell like? And will you ever be able to forget it?

Who wouldn't be wooed by these sonnets that nature whispers? These magical moments create the foundations for future scientists. I'm yet to meet an ecologist who entered the discipline because they wanted to attend more faculty meetings or wrestle the photocopier.

No, their journey begins with cheap tents, charged batteries, many lost pens and a lust to observe the brightness and quietness of nature. Most things in that list don't change.

Field work is, I think, the environmental equivalent of the debutante ball; your student is presented to the world and their adult life is just about to begin. Awkward photos are guaranteed.

I've seen addicts (ecologists) with the code blue wobbles associated with skipping a field season. It becomes part of the rhythm of their existence, just like the passing of seasons or university terms. It's often a necessary break from 'real life'.

Is it a marathon? A labour? Or is it, as people who have never gone out into the field may think, simply a holiday?

It's none of these and all of them. Sometimes in quick succession. It's agony and ecstasy. Pain *and* gains.

I've always imagined the world as though I was a satellite. While I orbit outer space, I can observe the globe as a whole. Then I can zoom in on a continent from way above, then a city, a suburb, a garden, a plant, the humus, the soil, the bacteria in between the grains. Then a giant worm makes the earth crackle as it goes past and I can see the ridges on its body.

The only other people I've found who hold in their brains such contrasting scales of the macro and micro world, with both stark realism *and* unabashed whimsy, are ecologists.

Regardless of how much of their physical time is spent in the field, much of their mental time is spent in the field. They have a direct line to Mother Nature, praying at the altar of field work, committed and devoted.

Of course it's not all sunsets, fresh air, new species and enchanting sounds. It's more often sweat, dirt, bites, less than ideal hygiene and far from ideal toilets. But everyone still yearns to be in the field. Everyone knows where they're going next, even if they don't know when they're going to get there. Because it affirms who we are in the world.

Never stop looking under bushes.

Dr Ann Jones
Presenter, ABC Science

CONTENTS

PREFACE

It's twilight on the stony gibber plains of the southern Kati Thanda-Lake Eyre Basin. The desert is quiet, except for the flies as they buzz around my face and attempt to investigate the inner reaches of my nostrils. The crickets have begun to chirp. I spot a single raven flying northward to an unknown destination and hear its wings beating as it passes overhead. 'Git orf' I mutter, swatting flies with my mind. I dare not move, for fear of disturbing the tiny furred animal that I am here to observe.

Alone in a remote Australian desert, with mostly just flies for company, I was collecting data for my PhD. I wanted to learn more about the secrets of the vast fields of deep cracking clay soils found in parts of that region, and the animals and plants that relied on them. In particular, I wanted to understand how cattle grazing and rainfall influenced the dynamics of the local flora and fauna.

It was winter and, although the weather had been kind during the day, as the sun disappeared the temperature would drop sharply. I sat watching a stripe-faced dunnart, a small dasyurid (carnivorous marsupial) about the size of a mouse. This diminutive creature was the reason for my stiff aching limbs and my silent mental stoush with the flies. The dunnart, whom I had captured the previous night in a small metal box called an Elliott trap, was missing a minute section of her fur (thick and light grey with a darker undercoat), to which I'd attached a small radio-transmitter, before releasing her. I listened to the transmitter's signal, a soft persistent 'thwok', like a heartbeat. I hoped that this electromagnetic signal would help reveal some of the secrets of dunnart life on the cracking clay and gibber plains, including where she went after night fell (dunnarts are nocturnal) and which cracks in the soil she used to shelter in, perhaps hiding from larger predators like inland taipans, dingoes or barn owls.

As the sky turned from an unrelenting clear blue to a light violet hue, and as my foot fell asleep, I suspected that I was in for a long night, following this small nocturnal predator about the gibber plains. The dunnart began moving about furtively. Perhaps she was considering making a dash for freedom down a nearby crack in the soil? Instead, she resettled into her hiding place underneath a *Sclerolaena* shrub – a small bush covered in multitudes of tiny spiky fruit that poked at my backside every time I crouched too close ... The dunnart glared back at me balefully with dark liquid eyes. If she was any larger than a mouse, I'm sure I would have been toast.

***

Many years after this experience, I still read through my field notes from that time and smile. I recall the solitude of the desert and my excitement at tracking a dunnart for the first time. The memory is still vivid, recorded in one of the tattered and rather stained (with mouse urine, dunnart faeces, squashed mosquitoes and other detritus) field notebooks that I kept throughout my postgraduate studies. Those ageing notebooks, and the memories recorded in them, are precious to me and partly inspired this book. This book was also motivated by an unwavering belief in the critical importance of field work for conservation and science, and for helping us understand our place in the world. Exactly why do ecologists spend so much time away from the comforts of home, in often inhospitable (or at least vaguely uncomfortable) places, doing things that might appear to any 'normal' person to be somewhat ridiculous?

Having grown up in Central Australia, on the edge of the Tanami Desert among red sand dunes and spinifex, I am most at ease in those sparse and arid places where life clings on perilously, but where plants, animals and people persist (and even thrive) despite little rain and brain-melting summer heat. I'm not particularly spiritual nor religious, but I have an unshakeable belief in plants, animals, fungi, soil, rocks and the complex web of life and evolution that ultimately links all living organisms (and those that have ever lived). My faith in these phenomena has kept me steady in a sometimes puzzling human world; my choice to remain close to them has helped forge my career as a conservation biologist.

Field work is the foundation of the ecological sciences. Field-based scientists go out into the field and collect often hard-won data (e.g. specimens, counts of animals or plants, behavioural observations and even animals used in the laboratory) to 'feed' the laboratory work or the computer models that are analogues of natural systems and are often used to guide decisions in conservation and environmental management. Sadly, field work is a flailing, if not dying, part of the ecological and biological sciences. Scarce funding opportunities, cost-saving measures, and perhaps a form of risk paralysis in the organisations that host field biologists have made it harder than ever to 'do' and teach ecological field work.

The importance of field work and field studies to building the skills, knowledge and capacity of students – the next generation of scientists and conservationists – is undeniable, and yet opportunities to learn field-based research skills are dwindling across the education sector worldwide. Inexplicably, this pattern is happening despite global recognition that biodiversity, wild places and the organisms and natural processes that live in, depend on and support them, are declining. Indeed, the origins of ecology (the study of organisms and how they interact with the environment around them) are rooted firmly in data collected through field work and natural history observations. These data inform the models and meta-population analyses that are a mainstay of contemporary ecology and conservation biology.

It is curiosity, persistence and a passion for conservation that motivates field researchers to explore the exquisite beauty of a tiny dasyurid who lives in the desert, hunting invertebrates under the cover of night, and moving across vast areas to take advantage of the desert's sporadic rainfall. Or be humbled by the realisation that, human or otherwise, we are all ultimately potential prey for another species (see Neil Jordan and Robert Heinsohn's chapters on working around large predators). Or, if they are lucky, feel the euphoria of unexpectedly discovering an unknown population of a threatened species (see Jodi Rowley's chapter on finding lost frogs). Or perhaps learn the importance of team work and leadership from a beautiful dog while counting carnivorous plants (see Laura Skate's chapter about a dog named

Bonnie). Some field biologists seek to understand a species and the threats it faces with intricate detail, by spending much of a lifetime studying it in the wild (see Andy Derocher's chapter on polar bear research). How can we *really* know that a species or system is in trouble unless we go into the field, and look?

Usually, data collected through a field study are analysed and then published in technical scientific documents, such as peer-reviewed journals, and shared among peers at conferences. Publishing your data is typically regarded as the pinnacle of science (perhaps misguidedly). However, behind these papers and their sometimes complex analyses lie tremendous tales of challenging, hilarious and occasionally terrifying experiences of researchers who asked a question about the natural world and set out to answer it. These experiences are the 'human' side of research and can be the catalyst for changed perspectives, inward reflection and even improvements to our understanding of ecological theory.

The authors and I have attempted to transfer these experiences from our memories and field notebooks into this book, in the process examining why we are driven to do the work we do and what we learn by doing it. We hope that these stories will inspire you to challenge yourself, take risks and look inwards from time-to-time (while being sure to watch out for any hyaenas that may be following you).

ABOUT THE EDITOR

Dr Helen P. Waudby is an Adjunct Research Fellow with the Gulbali Institute at Charles Sturt University in New South Wales, and a conservation biologist with the New South Wales Government. She has worked in wildlife research and conservation for over 20 years, spending much of that time in the field, in places ranging from wetlands to mallee environments to the great stony deserts of northern South Australia.

ACKNOWLEDGEMENTS

First, thank you to my partner, Matthew Gill, for your unwavering support and field assistance over the years, including helping me capture and count thousands of ticks when no-one else would, digging endless pitfall traps, and cooking all those dinners in the field. I continue to be amazed by the gourmet meals you can prepare when all that's available is a slightly mouldy carrot and tin of tomatoes. Those desert sunsets and sunrises are always more vibrant when you're there.

I must acknowledge the insightful, fascinating and often hilarious contributions of the authors who agreed to be a part of this book. I really appreciate their hard work, support, patience and good humour. Thanks to them, preparing this book has been a wonderful and enlightening experience.

Thank you to friends and colleagues who have inspired me (and continue to do so) over the years with their own tales from the field, and their great love for and knowledge of the natural world. We are all field naturalists at heart. I am also indebted to the volunteers and students who have helped me in the field. Many of them, quite literally, shed blood, sweat and tears in the name of conservation and wildlife research, and have become life-long friends. People's true natures are revealed during long hours of field work, whether it be lurking around mosquito-infested wetlands at all hours of the night looking for frogs or digging vehicles out of bogs for the umpteenth time. You are all legends in my view.

My gratitude to CSIRO Publishing, including Mark Hamilton, Briana Melideo, Eloise Moir-Ford and Tracey Kudis, for their advice, support and helpful comments on preparing the manuscript. Sincere thanks to Brad Smith for his advice and encouragement in proposing and preparing this book. I am also indebted to Brad and to David Parker for proofreading and providing useful and pragmatic comments on the preface, and to Jenny Sedgwick for her helpful comments on several early chapters. My sincere gratitude to S. Topa Petit who has been a significant mentor over the years and was the first person to show me that conservation biology is a real career.

> I went to the woods because I wished to live deliberately, to front only the essential facts of life, and see if I could not learn what it had to teach, and not, when I came to die, discover that I had not lived. (Henry David Thoreau)

LIST OF CONTRIBUTORS

Tracy Ainsworth
Associate Professor, School of Biological, Earth and Environmental Sciences, University of New South Wales.
tracy.ainsworth@unsw.edu.au

Wayne Boardman
Associate Professor in Wildlife, Biodiversity and Ecosystem Health, School of Animal and Veterinary Sciences, University of Adelaide.
wayne.boardman@adelaide.edu.au

Isabel Castro
Professor, School of Natural Sciences, Massey University.
i.c.castro@massey.ac.nz

Pam Catcheside
Honorary Research Associate, State Herbarium of South Australia; Adjunct Lecturer, College of Science and Engineering, Flinders University.
Pam.Catcheside@sa.gov.au

Andrew E. Derocher
Professor, Department of Biological Sciences, University of Alberta.
derocher@ualberta.ca

Christopher R. Dickman
Professor, School of Life and Environmental Sciences, The University of Sydney.
chris.dickman@sydney.edu.au

Leonardo Guida
Shark Conservation Lead, Australian Marine Conservation Society.
leoguida@amcs.org.au

Terrah Guymala
Board Director, Warddeken Land Management.

David G. Hamilton
Adjunct Researcher, School of Natural Sciences, University of Tasmania;
Conservation Ecologist, Tasmanian Land Conservancy.
d.g.hamilton@utas.edu.au

Robert Heinsohn
Professor, Fenner School of Environment and Society, Australian National
University.
Robert.Heinsohn@anu.edu.au

Dieter F. Hochuli
Professor, School of Life and Environmental Sciences, The University of
Sydney.
dieter.hochuli@sydney.edu.au

Jo Isaac
Principal Ecologist, Ecology & Restoration Australia.
jo.isaac@eraus.com.au

Neil R. Jordan
Senior Lecturer, Centre for Ecosystem Science, School of Biological, Earth
and Environmental Sciences, University of New South Wales; Taronga
Conservation Society Australia; Botswana Predator Conservation.
neil.jordan@unsw.edu.au

Laura Kojima
Graduate Research Fellow, University of California, Davis.
lvkojima@ucdavis.edu

Jessica Marsh
Adjunct Research Fellow (Arachnology), Harry Butler Institute, Murdoch
University.
jess.marsh@murdoch.edu.au

Lydia McLean
Field Coordinator and Community Engagement Coordinator, Kea Conservation Trust.
lydia.mclean@keaconservation.nz

Cara Penton
Ecological Manager, Warddeken Land Management; Research Fellow, Charles Darwin University.
cara.penton@cdu.edu.au

Euan G. Ritchie
Professor in Wildlife Ecology and Conservation, School of Life and Environmental Sciences, Deakin University.
e.ritchie@deakin.edu.au

Erika M. Roper
Ecologist and Environmental Science Communicator.
erikamroper@gmail.com

Jodi J. L. Rowley
Curator of Amphibian & Reptile Conservation Biology, Australian Museum and University of New South Wales.
Jodi.Rowley@australian.museum

Manu E. Saunders
Senior Lecturer in Ecology and Biology, School of Environmental and Rural Science, University of New England.
Manu.Saunders@une.edu.au

Laura M. Skates
Botanist and Environmental Science Communicator.
skates.laura@gmail.com

Bradley P. Smith
Senior Lecturer in Psychology, College of Psychology, School of Health, Medical and Applied Sciences, Central Queensland University.
b.p.smith@cqu.edu.au

Abi T. Vanak
Director, Centre for Policy Design and Senior Fellow (Professor), Ashoka Trust for Research in Ecology and the Environment (ATREE).
avanak@atree.org

David M. Watson
Professor of Ecology, School of Agricultural, Environmental and Veterinary Sciences, Gulbali Institute, Charles Sturt University.
dwatson@csu.edu.au

Helen P. Waudby
Adjunct Research Fellow, Gulbali Institute, Charles Sturt University.
hpwaudby@gmail.com

1

Ngarridjarrkbolknahnan kunred: Looking after Country together in West Arnhem Land

Cara Penton, Terrah Guymala and Warddeken rangers

I t's early morning in northern Australia, during the wet season. The darkness slowly gives way to light as the sun rises over the Mankungdjang (native honey dreaming) estate, nestled in the West Arnhem Land Plateau, ~500 kilometres east of Darwin. The manbordokorr (stringybark) and mandadjek (grevillea) trees cast shadows as the camp dogs stir, sensing that it's time for breakfast at Kabulwarnamyo, the community founded by Bininj Nawarddeken's unwavering desire to return to their ancestral lands. The morning air is slightly cooler than the usual mid-30s°C we normally experience at this time of the year, but humidity is already high. As the kettle starts to whistle, the dogs rustle the leaf litter as they try to creep to the communal outdoor dining area, knowing that balanda and bininj are about to start their day.

Just before 7:00 am, we're ready to leave Kabulwarnamyo. We pack gear under the seats of the helicopter and strap fire rakes and tubs to the litter rack. We are travelling by helicopter, but it can still take 40 minutes to get to some of our monitoring sites, as we travel up to 85 kilometres away from Kabulwarnamyo to survey biodiversity on Indigenous clan estates. Each day we're a team of four; this morning the team consists of myself, Zecchaeus and Frankie, both Warddeken rangers, and the helicopter pilot Slarky.

The helicopter lands at a rocky outcrop, the ground damp with the lingering scent of overnight rain. The rocks glisten in deep red hues; patches of dark soil have collected in their recesses over decades. At the camera site my job is to collect the vegetation data. We conduct a rapid vegetation assessment, clearing the area of spinifex as Zecchaeus and Frankie set up the camera, attaching a peanut butter bait station to lure small mammals in front of it. Both yawurrinj (young men) have been setting wildlife monitoring cameras for years, which is why they have the responsibility of camera set-up today. Meanwhile, I measure out the vegetation plot, creating a square around the camera tree. I record grass cover, shrub cover and tree measurements, and assess the diversity of vegetation structure as I walk around the plot. I usually end up walking around the camera tree four to five times. My average step count in March is the highest for the entire year. Zecchaeus and Frankie chat away in Bininj kunwok, pointing towards Gunbalayna, a township in West Arnhem Land which is at least 50 kilometres away. Zecchaeus starts pointing in another direction, talking about Darwin, and then mentions Sydney; Frankie pauses for a moment and flicks his hand south-east. Frankie's never been to Sydney, but says he'd like to visit one day. At this stage, I've walked around the plot enough times to completely disorientate myself from the position of the helicopter let alone know what direction Darwin is in relation to our current location. I don't know whether I've become more relaxed over the years or was just always terrible with directions; Bininj Nawarddeken always know where they are. I suspect that if you ask anyone who I've given directions to (in person, in a vehicle or in a helicopter) they'll shake their head and tell you I've always been hopeless! Despite my poor sense of direction I've never been seriously lost, and certainly never when working with Bininj Nawarddeken.

I started with Warddeken in 2021 as their ecology officer, after finishing my doctorate studies on the Tiwi Islands. Having just submitted my thesis, I travelled to Manmoyi outstation and was interviewed by a collective of 10 senior Bininj from across the Warddeken ranger bases, for a role that primarily involved coordinating and facilitating the long-term biodiversity monitoring program the organisation had established in 2017. Concerned with the dramatic decline of native species across northern Australian, and observing these declines in the Warddeken Indigenous Protected Area (IPA), Warddeken rangers, with support of and co-design with Traditional Landowners, established a network of 120 monitoring sites across the IPA. Twenty Traditional clan estates participate in the program. For Bininj Nawarddeken, many species have a special cultural significance – their loss is not only a loss to biodiversity but can also mean a loss of Bininj knowledge, as customary obligations can no longer be performed and passed on. In the beginning, each year 60 sites were surveyed with five cameras for five weeks. To re-evaluate the sites, and in line with long-term systematic surveys across the Top End (as the Northern Territory is affectionately known), we now sample about 40 sites per year. This schedule means it takes three years to survey all 120 sites.

In 2023, we were sampling the sites for the third time since they had been established. Committing to long-term biodiversity monitoring is the only way to understand the impacts of threats and land management activities on key species of cultural and conservation concern over time. This information assists Traditional Landowners in their Caring for Country decision-making. The program's priority is to deliver the work with meaningful engagement and employment of Traditional Landowners and Indigenous rangers. Doing so means working across several outstations – Kabulwarnmyo, Manmoyi and Mamadawerre – and supporting travel for key Traditional Landowners from Gunbalanya, Maningrida and Jabiru. My role is primarily to facilitate a means for people to connect with Country, to understand the biodiversity present on estates and explore how we may conserve it together.

As we meander back through the bush towards the helicopter, we no longer need the GPS to find the camera trees and I shove it down my pocket. Frankie navigates through the sandstone rocks and spinifex, and I wander behind. I observe the sparse *Corymbia* trees with speckled bark and light patchy canopies dotting the landscape. This area hasn't seen fire for several years, so dense clumps of seeding spinifex (flowering spindly shrubs with pops of purple) and creeping reclinata resembling flattened cacti thrive. All of these species I've recorded through the vegetation plots. I'm hoping that in this kind of habitat we'll record mulbbu like Arnhem rock-rats, sandstone antechinus, maybe a darrara (Kimberley rock monitor) or a bulurr (ridge-tailed rock monitor) if we're lucky. We won't know for a few months; after leaving these cameras out for at least five weeks we'll collect them, and the images will be distributed between the bases for rangers to identify the species in a bilingual database.

The heat is intense and we feel ourselves baking on the rocks, sweat dripping down our faces. We find a shaded spot for lunch and sit down together, sharing food and stories, enjoying the company and the beauty of the rugged landscape around us. Near us is some bim (rock art) on the kuwardde (stone), the bright white paintings starkly contrasting against the orange sandstone. Frankie explains that these bim are the stories of Bininj Nawarddeken, the local Indigenous people, told over thousands of years. Some were painted by people, while others were placed by the spirits. We see depictions of djabbo (northern quolls), djankerrk (thylacines) and namarnkol (barramundi), all coming together to form a gallery of ancient art and storytelling. A motif of several birds is lined up in a row at the bottom of the gallery. I ask Frankie what he thinks the birds are and he jokingly exclaims, 'I think penguins for sure!' This quip catches me by surprise and cracks me up. Whether the bird has a secret story, or Frankie used the opportunity to tell a killer joke, I'll never know.

As we approach the next site, a woodland near a spring, I am reminded of my previous visit to this place two years ago. The woodland is filled with grey stringybark and white paperbark trees, adorned with fluffy pompoms of white flowers. The grass stands tall, sticking straight up because of the rain from the wet season, and it's a pale pastel green colour, dotted with vibrant orange pea flowers. The memory of landing in the spring on our previous trip stays with me. I open the door of the helicopter to exit, immediately notice the smell of smoke and exchange a quick glance with the pilot. Without hesitation, we jump back into the chopper and lift off. The catching of the dry grass could start an engine fire if we stayed for too long. Luckily, we had noticed another landing spot a few hundred metres away.

As the helicopter starts to lower for the second time, I spot a large nalangak (frilled-neck lizard) running through the grass and dashing up a short yilbinj (paperbark tree) right next to our landing spot. The lizard is clearly unhappy about being disturbed by the noise and wind from the helicopter, and opens its frill in defiance. However, with the force of the wind from the helicopter, the lizard ends up looking like an umbrella being forced inside out in a storm and it loses its grip on the tree trunk, swinging out to the side. Despite its valiant efforts to regain its balance, the lizard repeats the same mistake and pushes its frill out again, with the same result.

I can't help but burst into laughter, and I point out the comical scene to Zecchaeus and Frankie, who also chuckle, exclaiming 'warre (poor fella)'. As soon as we land, I exit the chopper and run over to the tree where the nalangak has now perched on a low branch. Without the disturbance of the helicopter, the lizard is now fully expressing its distaste, hissing, swaying and displaying its frill in all its glory. With cane toads decimating populations of larger reptiles, such as the nalangak, burarr (water monitors) and maddjurn (black-headed pythons), it's heartening to see large adults thriving.

As we reach our third camera site, Frankie's eyes spot a tiny native honeybee nest on the ground. He explains that this bee and the honey it creates hold special significance to him as it is part of his djang (his dreaming). This type of honeybee nest is built underground with only a small opening to the surface, about the size of a 50 cent coin. He and Zecchaeus talk among themselves in Kunwinjku, discussing what to do with the nest. They decide it's okay to set up the camera at the tree but they clear the area around it carefully, mindful of the bee nest. Frankie takes extra care not to disturb the bees, clearing a fire break around the camera tree as he wants to protect the native honeybee nest from future fire. He sets up the camera facing west instead of south as the usual location of the bait station would impact the bees, ensuring that their nest and surrounding habitat is safe.

Once the wildlife monitoring camera is set up to his satisfaction, Frankie takes out his phone and captures photos of the nest and bees, documenting the moment. As scientists we often strive for standardisation, uniformity and repetitiveness. Replication is the backbone of science but as ecologists and people sometimes best-practice is a bit of variance. Returning to the helicopter, we collect yellow rocks along the river creek. The yawurrinj collect these rocks to grind into traditional paint with other Bininj back in community to paint bim on canvas and madjdjul (strips pulled off a stringybark tree). The art pieces will be given to pilots returning to Gunbalanya and Jabiru to pass on to the art centres, supporting the local Indigenous community's cultural practices and craftsmanship.

As we make our way back to the community in the helicopter, we navigate through the isolated storm clouds that have formed across the horizon. The cooler weather is a relief from the scorching heat, but it means that we have to contend with rain during our flight. The landscape below is vibrant green with pulsing rivers and thundering waterfalls because it is the wet season. The land will soon transform into a brittle yellow place with dry tributaries and exposed sandstone as the dry season sets in.

The rain disappears behind us and the horizon is once again clear. In the late afternoon with the dipping of the sun, the light is softer and the savanna glints, enticing you to stay. 'Every day's a diamond, isn't it?' the chopper pilot says, using a bit of playful sarcasm, mimicking the phrase often used to express appreciation for the simple joys in life. The yowurrinj look up, making eye contact, one gives him a thumbs-up and the other laughs, appreciating Slarky's attempt at banter. I chuckle as I don't think the joke landed as Slarkly anticipated but, despite its inflection, he really wasn't being sarcastic. In the remote and wild beauty of the Top End, every day can feel precious, filled with unique experiences and unexpected moments.

Despite the challenges and risks across the landscape, nganabbarru (swamp buffalo), kinga (crocodiles) and nakurl (knock 'em down storms), the opportunity to witness one of the most unique and intricate landscapes in the world is a privilege. This privilege doesn't come without dedicating time and much of yourself to the land, but with it comes the opportunity to share knowledge and join together to gain new knowledge of the environment we live in. The last of the sun glints through the savanna tree edges. The helicopter lowers past the familiar cave and spring that let us know we're home. It's the end of another day and tomorrow will be a different crew, visiting a different place, with unknown experiences awaiting. After all, Caring for Country is forever.

What is most special about the Warddeken biodiversity monitoring program to me is that it's a profound example of how traditional knowledge and practices weave through modern conservation efforts, honouring the land and its significance to Bininj Nawarddeken. Around 80% of the world's biodiversity exists on Indigenous estates and it is only by supporting Indigenous people and local communities, who are connected to and know the land best, that it will be looked after for future generations.

***

We've got to find a better way of finding mayh (animals) that are missing. We started with the old-style ways (Elliott and pitfall traps) when we were looking at the area where mayh might live. We know this Country because we used to walk around it and see them a lot, but when all these things started slowly disappearing we said, 'Oh, something bad is happening here'. Those times when our people left this Country and met up with the missionaries, all this Country was empty, and the wildfire was going through. After we returned home, we had cats. They went wild and spread everywhere. We didn't know that they meant more danger for our native animals. Then pigs and buffalo came, and destroyed our good habitat for our little mayh, waterways and all those sorts of things. That made the Country really sick, and it was slowly, slowly disappearing. When we reached the late 1990s and early 2000s, we started to see more bad signs. Old people could see that something was happening here. I reckon it was slowly dying when we returned, that's what the old people thought, but we didn't know for sure until now.

When we first started to do big surveys, we knew this area and we went to find just a little evidence of mayh, like droppings from little animals or tracks. We said, 'Okay, let's put traps here and try to find if we can look in this area'. We found nothing, we didn't see anything, just toads and some lizards. Most of our pitfall traps were full of toads and snakes. The cameras were good because they're there for five weeks and we see all the mayh that get caught on camera when they walk past. But what next? We want to see mayh numbers coming back and have more monitoring sites, so we can start checking if numbers are going up or down. We started managing feral mayh and we were doing good fire management. We saw all the good signs to show that our management was improving, and we've seen the emus come back. We're still worried about other mayh like djabbo (northern quoll), yok (northern brown bandicoot) and nawaran (Oenpelli python).

When we first started looking for all these mayh, we couldn't see them (or we couldn't find them), but today when we look for them, they're still around. For us, it's sort of like things are coming back to normal for bininj and mayh in this Country. All these things that were missing are coming back like a missing parcel and us mob are still here. The old people told us our songs, the Morrdjdjanjno, for nawaran, for djabbo, bakkadjdji (black-footed tree-rat), nabarlek and badbong (short-eared rock wallaby). We are losing this knowledge as well, which also might be why mayh are disappearing. So, we need to start performing our ceremonies and songs again. They are connected – this land and the song that will bring mayh back to the landscape again.

Acknowledgements

Warddeken would like to thank all the Traditional Owners and rangers who have given permission and assisted with ecological monitoring over the last seven years. We want to acknowledge our Bininj Nawarddeken professors, those we have with us and those whose spirits have returned home to the kuwarddewardde.

Bininj kunwok dictionary/glossary

Badbong – short-eared rock wallaby

Balabbala – platform or bush housing

Balanda – non-Aboriginal person

Bakkadjdji – black-footed tree-rat

Bim – rock art

Bininj – Aboriginal person or man

Bininj kuwok – Aboriginal language which includes six dialects in the West Arnhem Land region of the Northern Territory

Bininj manbolh – traditional walking route

Bininj Nawarddeken – Aboriginal person of the stone country

Bulurr – ridge-tailed rock monitor

Burarr – water monitor

Daluk – women

Darrara – Kimberley rock monitor

Djabbo – northern quoll

Djang – place of significance where the essence of a totemic being resides, or a reference to the totemic being itself

Djankerrk – thylacine

Kakbi – north

Kinga – saltwater crocodile

Koyek – east

Kunwinjku – dialect of Bininj kunwok

Kuwardde – stone, rock

Kuwarddewardde – stone country, also referred to as Warddeken

Maddjurn – black-headed python

Madjdjul – outer strips of bark from *Eucalyptus tetradonta*

Manbinik – *Allosyncarpia ternata*, also referred to as anbinik

Manbordokorr – stringybark tree or *Eucalyptus tetrodonta*

Mandadjek – fern-leaved grevillea or *Grevillea pteridfolia*

Mankungdjang – native honey dreaming

Mayh – can refer to any animal

Morrdjdjanjno – songs performed as part of animal increase rituals or ceremonies

Mulbbu – general term for various small dasyurid and similar sized small mammals

Nabarlek – narbarlek rock wallaby (Top End)

Nakurl – knock 'em down storms, which are the very last storms of the wet season, generally occurring in March–April

Nalangak – frilled-neck lizard, also referred to as kurmdamen or danngarr

Namarnkol – barramundi

Nawaran – Oenpelli python

Nganabbarru – swamp buffalo

Warre – poor thing, too bad, oh dear

Yawurrinj – young men, boys

Yilbinj – paperbark tree species

Yok – northern brown bandicoot

2

Avoiding arrest while chasing hedgehogs

Christopher R. Dickman

Some biologists get their kicks by working on big fierce animals. Others prefer to get up close and personal with assorted bugs, beetles or other small beasts, while still others – unaccountably – enjoy modelling nature from their computers. My own fascination lies with mammals that eat insects. A yet-undiscovered gene may explain this strange and arcane behaviour, for I have been obsessed with studying these unobtrusive creatures since first seeing hedgehogs in my suburban London backyard as a boy. I was immediately fascinated by these snorting, spiky animals that would swagger into the garden and displace the family cat with impunity, while he was feeding at his own food bowl. At school, I was fortunate to have teachers who would take kids on jaunts to trap small mammals before classes began. These excursions were my first exposure to another insectivorous group of mammals, shrews. These experiences continued when I started an Honours project on shrews in conifer plantations in Wales, and later a PhD project at the Australian National University, centred on some of the most intriguing of all marsupials: so-called 'hedgehog-equivalents' in the genus *Antechinus*.*

While still finalising my PhD thesis in Canberra in 1982, my student visa expired, necessitating a return to England. However, before the guillotine fell, I enquired about gaining an honorary position at the University of Oxford so that I could complete my thesis. I was pleasantly surprised (actually, stunned) when the answer came back as 'Yes, we can find room for you.' My wife Carol and I packed our bags. Our benefactor

---

* Hedgehogs and antechinuses look as different from each other as cats and dogs. When antechinuses were first described from animals that had been dragged out of preservative solution, I wonder if the spikes of their clumped wet fur reminded early colonial zoologists of hedgehogs back in the 'old country'.

was Dr David Macdonald, then the head of the 'Oxford Foxlot' and now Professor and Founder of the Wildlife Conservation Research Unit at the University of Oxford. David's dynamism and charisma had attracted a large group of students who were at various stages of their doctoral research. Local students were studying the behaviour of red foxes, cats and badgers in the environs of Oxford itself, and others were pursuing carnivores in the wilds of Iceland, South America and Africa. In addition to completing my PhD thesis, my task in David's group was to answer a question that had eluded the Foxlot up to that time: what was the prey base that supported populations of foxes, cats and badgers within the City of Oxford and its immediate surrounds? In other words, what were they all eating? It seemed to me that all three carnivores would be hunting small live prey like rodents, shrews and even hedgehogs. To confirm my suspicions, I completed surveys to find out where these small and mid-sized mammals occurred. I found that discarded KFC® boxes, fish and chip wrappings and other leftovers from takeaway meals also played a role in supporting urban foxes, but that is another story. It made the analysis of fox scats more entertaining than it might have been otherwise ...

Keen to start my new project, I set up sites where I could trap small mammals in different habitats within and just outside the City of Oxford. These sites varied from people's backyards to patches of scrub, long grass, orchard and forest. Such was the knowledge and reputation of the Foxlot among the city's residents that, as a *bona fide* Foxlot member, it guaranteed me access almost everywhere. With this universal passport, I ended up working in the backyard of the home that had once belonged to J.R.R. Tolkien, the expansive grounds surrounding the house of Sir Isaiah and Lady Berlin, and patches of forest associated with several colleges that had remained almost untouched for centuries. Surveys in these disparate habitats revealed 20 species of mammals, with some species hanging on in isolated and tiny patches of only 0.2 hectares. An important finding was that some mammals were ubiquitous. Hedgehogs and red foxes, in particular, lived in all habitat types and were present in 42 of the 50 different sites that I surveyed in the city area. The success of foxes seemed to come down to several factors. These wily creatures included in their diet most other small and mid-sized mammals found in the city, a range of birds, reptiles and invertebrates, fruits, and a bewilderingly diverse array of scavenged materials that included root vegetables, cooked meat bones, condoms, fish hooks, bottle tops and large quantities of garden compost. Foxes also made unobtrusive dens under garden sheds, usually emerging when residents were asleep or out. On rare occasions, though, proud adult foxes would emerge from cover with their cubs when people were about; the irresistible cuteness of the cubs ensured that Reynard's family was well fed by captivated householders for many days. Red foxes are clearly master con artists and experts at exploiting urban conditions, but how do we explain the success of urban hedgehogs?

Hedgehogs are distinctive small mammals and, in the UK, are considered harmless or even beneficial because they are believed to include garden slugs and snails in their diet. They also project a kind of spiky yet vulnerable charisma that beguiles many householders into giving them bread and milk and even placing small shelters in their backyards, so their 'pet' hedgehogs are safe and comfortable. The hard truth – that hedgehogs fare poorly by eating non-natural foods and seldom use the artificial cover that is provided for them – seems not to faze people but does beg the question of what food resources hedgehogs actually do eat in the urban environment they exploit so well. I had to find out!

As a first step, I collected hedgehog poo so that I could identify any food remains that had passed relatively intact through the gut. The poo-collecting was easy. Hedgehogs produce lots of distinctive droppings when they are active: they are 3–4 centimetres long, cylindrical, black, and resemble pieces of liquorice that have been carelessly scattered on the ground. If you are caught collecting hedgehog poo (a constant hazard in many sites in Oxford) you are likely to be considered eccentric but harmless; some helpful folks even humoured me by giving directions to places where hedgehog latrines could be found. The hard part is identifying the food remains in their poo. Although identification can be straightforward for hard-bodied prey such as snails, slaters or millipedes, where parts of the shell or exoskeleton pass intact, it is much more difficult to pick out soft-bodied morsels like larvae, slugs and earthworms that are mostly digested. Indeed, some droppings have the consistency of tar! Nonetheless, after sifting through a lot of hedgehog poo, it seemed to me that soft-bodied prey might comprise a large part of hedgehogs' diet. I just needed to find a better way to quantify it.

As a second step, I resolved to go back to basics and spy on hedgehogs as they snuffled about looking for food. In the 1980s no camera traps or other devices were available that could be used for remote spying activities. Watching animals forage meant just that: as the investigator, you had to be out in the field when the animals were active to observe them directly. Hedgehogs are nocturnal and easily spooked, so they must be observed quietly, from a safe distance and under artificial illumination. No problem. I equipped myself with camouflage clothing, rubber gumboots that didn't squeak when I moved, and a hand-held torch that I covered with red cellophane. At the time, red light was thought to be invisible to most mammals, making it an ideal way to watch animals but not disturb them. We know now that many mammals can detect red light but, as far as my hedgehogs were concerned, if they knew I was there it didn't seem to bother them.

The initial results were encouraging. Little hedgehogs were gung-ho and ran after anything that moved – provided it was smaller than they were. Slaters, spiders and various insects were on the menu, and on one occasion a young hedgehog pursued a garden frog for some distance before giving up the chase. Older larger animals appeared to be more circumspect about where they foraged and what they ate. They often eschewed small invertebrates and focused more attention on prey such as large beetles, snails and larvae. To make sure I was identifying these prey animals reliably, I became more high-tech and glued my red torch to a pair of cheap binoculars to get a close-up view of what the hedgehogs were eating (okay, I know – I use night vision goggles these days ...). The results began to reveal some unexpected patterns. Apart from what appeared to be a change in diet with ageing, hedgehogs seemed to target different types of prey in different habitats and when foraging in different environmental conditions. Damp nights seemed to drive hedgehogs into suburban gardens or onto short grass fields where rain had flushed grubs and worms to the surface. On dry nights, animals were more active in patches of scrub or forest.

Call me a fair-weather biologist, but most of my early nocturnal observations of hedgehogs were made on dry nights when conditions were good for visibility. The hedgehogs were easy to observe, and people on the streets were able to see me and appreciate what I was doing. Some folks would get quite close before seeing my red torch and camouflage fatigues, backing off quickly, whereas others noticed me at a distance and crossed to the other side of the road before running away. These helpful people obviously didn't want to disturb the hedgehogs. Thus encouraged, I decided it was time to expand the study and complete more sleuthing on wet nights. Doing so entailed waterproofing my torch and binoculars and purchasing a suitably camouflage-coloured plastic mackintosh (actually, it was bright yellow, but all I could afford at the time); my rubber gumboots were already fit for purpose. As it was summer in England I didn't have long to wait for a wet night. With a forecast of heavy rain for several hours, I set forth to record the behaviour of any hedgehogs that I could find.

My wet-night observations began unsuccessfully. I realised, belatedly, that hedgehogs may well have more sense than the people who study them – they generally seemed to stay snug and dry under shelter on nights of heavy rain. However, shortly after reaching this conclusion on one particularly miserable wet night, perhaps the fifth or sixth such night that I had endured, I spotted a large hedgehog splashing through puddles of water that had formed on the edge of a short grass playing field. Hallelujah – this was my chance! The hedgehog was behaving just as I had expected and was busily slurping up worms that were wriggling helplessly on the flooded grass. After eating half a dozen worms the hedgehog sauntered off towards the front garden of a nearby house and squeezed through a hole at the bottom of the paling fence to continue its supper inside the yard.

Now, a rational person would have taken this situation as a wet-night win and gone home to dry off and get some sleep. But it was a dark night and, in a distinctly irrational moment, I clambered over the paling fence to continue watching the hedgehog, wondering what new gustatory insights I might uncover. Had I been less intent on following my quarry as it fossicked about in a flower bed, I might have paid more attention to the householder who was peering at me through a chink in the curtains with a startled look on her face. The hedgehog meanwhile continued its fossicking, oblivious – like me – to the unfolding events.

Suddenly I heard a loud wailing and noticed that my thin red torch beam had been replaced with an all-pervasive red and blue flashing light. The hedgehog immediately froze. I turned around to find a strong beam of light in my face. I also froze.

'Sir, this is the police. Stop what you are doing. Come with us!'

I tried to explain that I was only looking at a hedgehog, as anyone might do on a stormy wet night, but alas by this time the hedgehog – my alibi – had shot through. The police evinced no interest at all in the fresh liquorice droppings that were strewn on the wet grass, despite my explanation that they provided irrefutable support for my story. I was taken to the station for questioning. The police had received a report from an alarmed householder, Mrs Hughes, that a strange man wearing a bright yellow mackintosh and rubber boots had scaled her fence and was wandering around in her front garden with a red torch. The poor lady had thought it especially odd that this man was in her garden on such a wet night. I wondered in passing if strange men in yellow mackintoshes and rubber boots cavorted in her garden on dry nights, but didn't volunteer that thought to the police. I was in enough trouble already.

A young constable took it upon himself to question me and, after establishing that I was neither drunk nor high on any illicit substance, became more and more amused by my story.

'I see, sir. I completely understand that earthworms *would* be hard to find in hedgehog poo. So, as a dedicated scientist, you just *had* to go out on a stormy night in combat gear and see if hedgehogs would eat drowned worms. Tell us again about the red torch and binoculars.'

The police station was a small one, and most of the officers who were on duty that night seemed to find a reason to drop into the interview room and add their 6 cents to the proceedings. There were lots of wry smiles. My Foxlot passport eventually proved to be my stay-out-of-jail card later that evening, with my friend David confirming in a late-night phone call that I was indeed a member of his group. Of course, trespassing on someone's property is not to be condoned, and I was fortunate that both the police and the householder eventually saw the funny side of the incident. Mrs Hughes even invited to me watch hedgehogs in her garden on the next night it rained.

As a final step in my quest to discover the food resources used by hedgehogs, I collected animals that had been killed on roads in and around the Oxford city area. If the carcasses are reasonably fresh the last meal eaten by each animal remains in its stomach and is only partly digested, so the ingredients of that meal can be easily recognised. Dead hedgehogs can also be aged by counting bone rings in the jaw (one ring = one year of growth), so any age-related change in the diet can be characterised. Perfect! Of course, finding road-killed hedgehogs was a hit-and-miss task (pun intended) and scraping carcasses off busy roads was not pleasant. However, I did have help. My hedgehog escapades achieved early notoriety and often attracted media interest, allowing me to place requests in local newspapers and on radio for information about where people had come across dead hedgehogs. I had intended to follow up any information myself, but many Oxfordshire residents were so keen to help that they collected hedgehog carcasses themselves and posted them to me. Their dedication to the cause was admirable. If you have encountered a dead hedgehog, especially one that has been stewing for a few days in the sun, you will know that there is no worse stink on all of planet Earth. Cloying, foetid, sickly and pungent are adjectives that could be applied, but all fail to convey the true nausea-inducing depth and stickiness of the chemical fog that emanates from the corpse of a departed hedgehog. It is a little-known fact that wrapping a decomposing hedgehog in newspaper and posting it in a cardboard box amplifies the bouquet. I realised this fact in dismay very quickly upon opening several of these malodorous time-bombs, but not before receiving a barrage of complaints from mail sorters at the Oxford post office and a heads-up from my new friends in the local constabulary. Back on local radio, I thanked the many citizen scientists who had assisted with the project but drew a very firm line at receiving any more gifts in the post.

The scientific culmination of the hedgehog project was very satisfying, with one publication on how hedgehogs become more discerning hunters as they age, and several further publications on small mammals in the urban environment. Despite the occasional bumps, my personal journey was also very gratifying. I came to realise that progress in science is not an inexorable and linear march towards 'the truth', but often a winding and serendipitous journey with many chicanes and cul-de-sacs to navigate. Had David Macdonald not allowed me to join the Foxlot, this particular hedgehog story would not have unfolded and my own journey in life would have taken a different path. I also came to appreciate the key role that people play in all aspects of conservation and ecological research. Engaging people at the outset of a project and sharing research findings with them ensures that research and conservation goals are more likely to be understood and achieved. And such engagement may just keep you out of jail!

Acknowledgements

I thank David Macdonald for allowing me to join the Oxford Foxlot, Sir Richard Southwood for providing access to the resources of the then Department of Zoology at the University of Oxford, and Carol McKechnie for her support at all times.

3

The Big Roo Count: kangaroos, kids and calamity

Euan G. Ritchie

On a dark, rainy, icy-cold night, in the depths of winter and under the towering trees of outer Melbourne's Yarra Ranges, the final preparations for an epic research trip are underway. It's ironic because the trip in question aims to resurvey antilopine wallaroo populations (roughly 10 years on from my PhD research on the species) across tropical northern Australia, where far higher temperatures and brightly lit and open savanna habitats will be the norm.

'Screech, crunch!' The sound of metal on metal, under many tons of pressure. A vehicle taking part in a four-wheel drive training course accidentally ploughs into the back of my vehicle, causing significant damage and making us spin and slide down a very steep and muddy hill. I was attending the 4WD course as a refresher, as the work we had planned entailed long periods of driving on sometimes seriously sketchy roads. We were supposed to be leaving in just four days.

Does fieldwork *ever* go exactly to plan? Even the preparations can bring surprises.

***

Despite the vehicle collision (nobody was hurt) we managed to get away in the wee hours as planned. The first day we drove from Melbourne to Dubbo, which is more than 800 kilometres and about 10 hours of driving – a long day in the saddle. However, this was no ordinary trip, my wife Jen and I were also taking our two children, just four and seven years of age. Were we utterly mad? In a few days' time, when we reached north Queensland, we'd be spending most of our days working and camping in very remote locations. In almost all cases, we'd be without electricity, running water, showers or toilets (just bush dunnies), but we would have extraordinary experiences creating memories that will last a lifetime and that conventional schooling simply can't replicate.

After many days and long hours of driving, with lots of games and audio books to maintain everyone's sanity, we arrived at Springvale station. It was the first site to be surveyed on 'The Big Roo Count'. Springvale station, like many of the study sites we'd revisit, is a working cattle station. The property is located near Cooktown, at the beginning of the Cape York Peninsula in far northern Queensland. So how do you survey antilopine wallaroos, and what *is* an antilopine wallaroo, you might ask? Surveying antilopine wallaroos and other macropods (science jargon for 'big feet') involves driving really slowly (10–15 kilometres/hour) for roughly 5 kilometres, noting what species you see, and recording their numbers, behaviour and habitat use. Antilopine wallaroos are one of two large macropods, the other being the black wallaroo, only found in tropical northern Australia. They are roughly similar in size to the better-known eastern grey kangaroo and common wallaroo, but their colour and markings are quite different from those species.

Surveys for macropods are best started just before dawn and just before dusk, avoiding the hottest hours of the day between about 9:00 am and 4:00 pm, when roos will typically find a nice shady tree, clock off and sleep. Clever. This schedule means that roo surveys often require waking up at 5:00 am to grab a quick breakfast on the fly and driving from camp to the start of the survey, before dawn. Imagine planning to maintain this schedule for three months, with a four- and seven-year old, while also setting up and taking down an entire camp every two days ... Yes, Jen and I truly were mad. A couple of weeks into the trip, to my surprise I received a phone call from the university safety officer, asking if I was undertaking a research trip with two young children in remote Queensland? I answered 'yes', as per the paperwork I had submitted and had approved by another safety officer, before our departure. Perhaps the challenge we'd taken on had only just dawned on others!

The survey at Springvale went to plan. Tick. Next stop, Rinyirru (Lakefield) National Park. This spectacular region is located further up Cape York. Termite mounds, native palms and other wonders are in abundance, including crocodiles. It was hot and dusty. Everyone would have loved a refreshing swim, and awfully inviting, waterlily-covered billabongs were everywhere, taunting us. Like the quintessential mirage in the desert! However, thanks to the very real threat of crocodiles, swimming was definitely off the cards. So too was camping anywhere near waterholes, especially since small, naturally inquisitive, free-ranging, wildlife-loving children were involved.

As a tangent in this story, I had one of my most stressful field experiences in Rinyirru, several years before this trip. At the same study site, I was a little too cautious crossing a creek when returning home from an evening roo survey. My slow speed meant that we got well and truly bogged in thicky oozy mud – in the middle of a creek, possibly home to crocodiles, at night. Under my instruction, my volunteers shone torches from inside the vehicle onto the surface of the water, checking for even the slightest ripple that could indicate trouble and my possible imminent demise. I summoned the courage to get out of the vehicle, wade through the water, attach the hand winch to the front of the vehicle and a nearby tree, and winch us out. I'm obviously still alive to tell this tale, but the memory still makes me anxious.

Another site down, and nobody, including our children, got eaten by crocodiles. We did get bogged in Rinyirru, but luckily it was in deep sand in a dry creek bed (better than thick mud!) and we got out easily with a few well-placed tree branches and lots of shovelling, and without the threat of large person-eating reptiles. We celebrated with a refreshing ginger beer at a roadhouse as we drove further north, which took the edge off being covered in a thick layer of congealing dust, sand, sweat and grime. Field work in the tropics is a whole different sport and challenge from work in other climes. It's best done with people you know, who can tolerate the conditions – and the smells – with good humour.

After resurveying several sites we finally arrived at the Steve Irwin Wildlife Reserve. We were very far north, fewer than 200 kilometres from the very northern tip of Australia. The roo surveys went to plan but at a rare evening social event, organised by the property managers, Jen and I received an almighty scare.

Snakes are a part of field work in Australia, and plenty of highly venomous species are present on Cape York, including the king brown, coastal taipan and northern death adder. We'd been careful all trip, wearing long pants and boots and avoiding walking through long grass, but at the party we let our guard down. Our kids were excited to have company, including children, and they were happily running around playing. Then our daughter rushed up to us, saying that she'd been bitten by something. We realised that she'd been playing at the base of a tree, which was surrounded by plenty of leaf litter – absolute prime habitat for a death adder. She had two little red marks on her skin. We were a long way from medical assistance. We sat her down, calming her, while suppressing our own rising panic. If she *had* been bitten the situation was dire. We attended to the 'bite' site. Minutes passed and we continued monitoring her body and behaviour for any alarming or unusual signs. Nothing. More minutes passed, and still nothing. Eventually we were satisfied that she was okay. To this day we have no idea what actually bit or stung our daughter that night – more than likely it was a spider or an ant – but it's a field work experience that neither of us will forget.

Setting up and packing down camp every two days and driving long distances between sites is challenging enough, but it's made even more taxing by the condition of many of Cape York's infamous roads. Most of the roads we travelled along were unsealed, and in many cases, heavily corrugated and punctuated by large dust holes. If you've driven along such roads, you'll understand all too well the whole-of-body experience these sort of conditions entail. The vehicle, our bones and our internal organs were jolted and vibrated, sometimes for what felt like forever. Dust holes are particularly dangerous, as they can be hard to see and if hit at speed can cause serious mechanical damage or even cause a vehicle to roll. Luckily, I'd driven these roads many times before and understood what was required to drive safely in such conditions. I'm not going to lie – it was always a relief to exit the car at the end of the day, brains slightly scrambled by the constant jarring and hours of concentration on the road.

It would be remiss of me not to mention some of the truly special moments that we experienced. Because of the hours involved in roo surveys, you get to enjoy truly glorious sunrises and sunsets, rising and setting through the savanna woodlands. The array of colours, their subtle variations and the changing nature of light in the woodlands throughout the day are simply extraordinary. Arguably, we had the most special experience of this trip at Piccaninny Plains. Early one morning during a survey, and coincidentally on Jen's birthday, we were joined by a dingo. It was not scared of us, nor we of it, although we had educated our kids about respecting this apex predator. Dingoes are very common on Cape York. This wonderful animal calmly trotted beside our vehicle for several minutes as we surveyed the landscape for macropods and then slinked back off into the savanna, out of sight. Amazing.

After several weeks we were finished with Cape York and moved through study sites in the Einasleigh Uplands, a region west of Cairns and Townsville. Many of the same challenges were present, although we could now swim and bathe safely as we were well out of estuarine crocodile country. Freshwater crocodiles were present in these waterways, but they don't typically pose a threat to humans – if we respect them.

You might be wondering what we did between morning and afternoon roo surveys? On many days, we did vegetation surveys and school work. Recording the plant species and structure of vegetation was important for establishing habitat preferences of the local macropod species, including the antilopine wallaroo. Anyone who has done this work will appreciate how taxing, and at times monotonous and laborious, it can be. Yes, walking through savanna habitat is lovely and you're often rewarded with unexpected wildlife encounters, but endless hours of measuring the height and girth of trees, trying to identify one shrivelled brown grass species from another, estimating ground and canopy cover percentages, and other metrics, all in the hot tropical sun, can grind on your nerves. Add school work on top of that? Yep. As we'd taken our kids out of kindergarten and primary school, we wanted to ensure that they were keeping up with their peers and the curriculum. So we attempted to run our own bush camp classroom. Their studies included writing about our experiences, recording what animals we'd seen and in what numbers – I wish my maths classes had been like that when I was a kid! – and art classes. Our son painted a beautiful blue-winged kookaburra that had unfortunately died on a barbed wire fence. So many important conversations were sparked on this trip.

If this experience sounds like a lot to take on, well, it was. We had our rough days and moments. Long hours and days of working, back-to-back, often in very trying and oppressive conditions, led to tired minds and bodies, tantrums and sometimes family arguments. Also, having to change a tyre that's blown out on the trailer, while lying in the dust with the thermometer well north of 40°C, can test your patience. But some mishaps can be amusing. Like the time we talked up having 'the best sausages in town' for dinner. Most of our food came out of packets or cans, because it was hot and we had only a very small car fridge. When you've bought these sausages, driven for hours, set up camp and started cooking, it's rather disappointing to find that your gas stove is broken. This disappointment is compounded when you realise you can't cook the sausages on a fire, as it's the dry season (annual period of little to no rainfall that typifies the dry tropics) and you're working in a national park (no open fires allowed in the dry season). We certainly wanted to avoid starting a bushfire. What do you do? You heave the uncooked sausages that you've been salivating over into the long grass, as they won't stay fresh until you can get your stove fixed, and you joke about how a dingo will likely be eating better than us tonight.

Those months spent together as a family conducting research were truly special. The trials and tribulations made us appreciate the joyous and special moments even more. Our kids have seen remarkable landscapes and wildlife, including green pythons, death adders and estuarine crocodiles (from a safe distance), spotted cuscus, eclectus parrots, palm cockatoos and of course many antilopine wallaroos. They've also learnt what makes savanna ecosystems tick and what it takes to conduct field-based ecology. We learnt and grew as a family, together, something we are forever grateful for. If there is a lesson, it's that science and research *can* be done by families together, but you have to be well prepared yet flexible, creative, supportive and patient. I adhere to these principles as I now lead a research group and the *family* – of amazing students and colleagues – I'm privileged to work with.

Acknowledgements

We would like to acknowledge Australia's first scientists and the Traditional Owners and custodians of Country throughout north Queensland, and more broadly all First Nations peoples across this vast continent. We also thank everyone who supported our crowdfunded project, The Big Roo Count, and Deakin University.

4

In search of a kingdom

Pam Catcheside

N o, not a fairy-tale kingdom, although some of the inhabitants do have fairy-tale qualities. I mean the Kingdom of Fungi. This kingdom can often seem elusive, difficult and challenging. It is all of those things but, for those who delve into its secrets, it is immensely fascinating and rewarding. It is a huge kingdom, perhaps as many as 2.2–3.8 million species with only 148,000 yet described.

Sometimes I think I am daft to have taken on, in retirement, the task of investigating the fungi of South Australia. We don't know how many fungi we have here but are fortunate to have ~16,000 collections of Professor Sir John Burton Cleland in the State Herbarium of South Australia. Cleland collected fungi mostly between 1910 and 1935, documenting ~600 species in his handbook, *Toadstools and Mushrooms and Other Larger Fungi of South Australia*. I am an honorary research associate in the herbarium and add to its collections by going into the field to make my own.

So, how did my interest in fungi start? As an undergraduate student in London, part of my botany degree covered mycology, the study of fungi. We had an inspirational lecturer, Dr Mike Madelin, who introduced us to the beautiful, the often rather ugly and the sometimes sinister members of the kingdom. He explained the roles they play in the environment and how fungi can be classified into groups according to their relationships with other fungi. After graduation I taught biology in high schools for almost 30 years. On retirement I decided to renew my acquaintance with fungi. While still teaching, I and my husband David often went looking for fungi at weekends and during school holidays. I had Cleland's handbook of fungi, but few field guides to Australian fungi were available in the 1990s. I needed pictures! English field guides had pictures, but I was finding that lots of Australian fungi did not look like any of the pictures or match descriptions in those books. Then, I read that a conference was to be held in Melbourne in October 1996 at which a book series, *Fungi of Australia*, was to be launched and a new Society, the Australasian Mycological Society, would be officially incorporated. Luckily, it was school holidays so David and I hot-footed it to Melbourne and I revelled in being among Australian mycologists. In 1995 Dr Tom May, Principal Mycologist at the National Herbarium in Victoria, had initiated Fungimap, a national society to map and promote Australian fungi. Tom also organised a post-conference field trip to Marysville, Victoria, where we were able to meet him and other mycologists from Australia and New Zealand. This field trip was a wonderful introduction not only to Victorian fungi but to the Australian mycological community, a community that is ever open and generous to newcomers, whatever their level of mycological knowledge or experience. That event in Marysville gave me the confidence to go 'fungologising' in the bush on my own or with David and get to know the local fungi. The more I delved, the more my enthusiasm grew, and my high school students became used to seeing displays of mushrooms and other fungi spread around the laboratory. Perhaps my students were not equally enthusiastic, but they certainly looked out for fungi and brought specimens to me, designating me a 'Fun Guy' – my first experience of that joke.

But why fungi? It certainly helps to have had an excellent teacher, but I get tingles down my spine when I see bright orange discs fringed with long black hairs of eyelash fungi *Scutellinia scutellata* or an earthstar peering through the leaf litter. Many fungi have beautiful forms and colours, but it may be their strangeness, their difference from plants and animals, and their elusiveness that attract me. Some pop up one day but are gone the next, their fruit-bodies melting into and enriching the soil. However, it's also the folklore surrounding fungi that fascinates me. The ancient Greeks and Romans had no idea how fungi originated and speculated that fungi, especially truffles, sprouted after lightning strikes and thunderstorms. That idea can be entertained even today. High electrical voltages have been shown to stimulate the fruiting of shiitake mushrooms, and rain accompanying a storm promotes fungal growth. Other folklore surrounds fairy rings, which were thought to be meeting places for fairies and elves whose feet generated the rings as they danced.

We now know that a mycelium, composed of microscopic hair-like tubes called hyphae spreading in all directions, is formed when a fungal spore germinates. At its circular growing edge, the mycelium may extract water and nutrients from the soil, resulting in a ring of dead grass. It also may fertilise the soil by breaking down organic matter releasing small molecules that can be absorbed by plant roots, leading to a ring of darker green. Mushrooms, the fruit-bodies, often form at the edges of the circle.

I have a sympathy for the underdog, or perhaps I should say underfung. Fungi have been called the orphan or Cinderella kingdom and are often maligned as dangerous and poisonous things to be feared. It is true that some fungi, for instance the well-known death cap *Amanita phalloides*, can cause death if eaten, but relatively few of the possibly millions of fungal species are highly toxic.

I admire fungi for what they do – they are quiet achievers. Without them, the Earth as we know it would not exist. They play essential roles in all ecosystems. Many act as recyclers and others form mutually beneficial relationships with plants. Even the pathogenic and parasitic species play a part in the natural turnover of living organisms in an environment.

Many fungi form special relationships (known as mycorrhizas) with plants, and almost all plants need their fungal partners. *Myco* is Greek for fungus, *rhiza* for root. A tree needs large amounts of water and minerals to function and grow. Its root system usually does not extend much further than the canopy formed by its branches, limiting its ability to collect these essential materials. Hyphae of mycorrhizal fungi may extend tens of metres beyond plant roots and into small spaces in the soil denied to the thicker roots, enabling efficient mining of needed resources. The recyclers break down the wood or leaf litter on which they are growing into small molecules that plants can absorb and use. These fungi are called saprotrophs (*sapros* is Greek for rotten and *trophos* is Greek for nourishment) and are responsible for almost all the recycling that happens in the environment. Without them, the Earth would be piled with dead plant and animal debris. Some fungi are parasitic. It is true that many are great nuisances; for example, smuts and rusts can cause huge crop losses. Smut fungi grow as sooty black, powdery masses of spores. Many rusts infect members of the grass family such as wheat, barley, maize and oats. As their name suggests, they form rust or brown-coloured powdery patches on leaves, stems and flowers. If they appear on the green leaves and stems, they reduce the area that the plant has to photosynthesise, and it starves. However, some rust fungi can be useful – they kill or harm a pest plant. For example, blackberry rust *Phragmidium violaceum* causes dehydration and defoliation of invasive blackberries, leading to die-back. *Puccinia myrsiphylli* forms orange patches and can decrease the number of tubers, rhizome length and shoot mass in bridal creeper, a climbing plant that can festoon and smother native trees and shrubs.

So, when I go on a fungal hunt, I go with a respect for what they do. I make sure I select good specimens that are unaffected by insect or other damage and that represent all stages of development from young to mature, to ensure that all the characters necessary for identification will be there. I collect a number appropriate for their size: up to five for larger specimens and 30–40 for tiny ones. Each collection is photographed in the field. I record data on ephemeral characters such as texture and colour, the date, location and description of the site and associated plants. Each collection is given a unique number. That evening I spend four to five hours doing further descriptions, microscopy and preparation of collections. Our fungal season, from the first autumn rains in late April until late August, is followed by six to seven months of laboratory work, research and writing.

All my collections are deposited in the State Herbarium of South Australia. A herbarium is a place where plants, algae and fungi are stored. It is rather like a library where collections can be 'borrowed' by others undertaking research. Some of the larger herbaria have special sections called fungaria where only fungi are stored. Herbarium collections are important as they enable further research on species, revealing information such as the range of habitats it occurs in, when it produces fruit-bodies and its geographic distribution. DNA can be extracted from herbarium specimens and analysis of that can provide information on the evolution of fungal species.

My first full year of fungal-hunting was in 1999 after I retired from school teaching. I was fortunate to gain a grant from the Wildlife Conservation Fund, which covered the cost of accommodation and travel associated with collecting. I also had the invaluable support network of the Fungimap community which, though scattered across Australia, gathered for biennial conferences and collecting excursions (known as fungal forays) at various places around the country. Each year, armed with my scientific collecting permit, I and often David surveyed around 40 parks and reserves. As our knowledge grew, we concentrated on the richer and more diverse sites. In the early years we travelled 7,000 kilometres or more each year, but the more time we spent travelling, the less time was available for surveying. On one occasion, we decided to explore the Moon Plains near Coober Pedy, 850 kilometres north-west of Adelaide. We thought the bare lunar landscape might yield some interesting fungi but the round trip of 2,000 kilometres produced only a single common earthball *Pisolithus arhizus*, and that a very poor specimen.

We learnt to concentrate our efforts on surveying wetter, well vegetated areas where fungi are more abundant. One early focus was Stringybark Walking Trail in Deep Creek Conservation Park, a remnant stringybark forest about 85 kilometres south of Adelaide. We recognised it as a fungal hotspot and published our results in 2008. Over the years we have documented over 400 species of fungi there, ranging from the delicate blue 'pixie's parasol' *Mycena interrupta*, the black clubs of earth-tongues, small orange and black disc fungi to the large brackets of white punk *Laetiporus portentosus* attached to the trunks of messmate stringybark *Eucalyptus obliqua*. These brackets are often tunnelled by insect grubs, causing the dry corky fungus to fragment into small pieces of white polystyrene-like material. On several occasions, I have been able to reassure those accompanying me that the mess is not due to careless members of the public disposing of their rubbish.

One of my favourite places is Flinders Chase National Park at the western end of Kangaroo Island in South Australia. It has large areas of remnant forest and woodland and is usually wetter than many places on the South Australian mainland. So far, the Chase has yielded collections of ~750 fungal species ranging from mushrooms, jelly, bracket and coral fungi to disc fungi. Some of the disc fungi are 'fire fungi', which fruit only after fire. Many more are yet to be discovered and described – park rangers frequently send me photographs of fungi, some of which I have not seen before.

Fungi can be found in the desert and semi-arid areas of South Australia, and have adaptations for coping with dry conditions. For example, the dry skin and spherical shape of puffballs help to prevent them drying out; a sphere has the smallest surface area to volume ratio, meaning that they lose less water than a mushroom with its more complex shape. Puffballs have a central small hole (a stoma) through which their powdery spores are dispersed. If you prod a puffball, you will see a small smoke-like cloud puffing out of its mouth. Some puffballs sit close to the ground, others have stalks giving them greater height to take advantage of wind. Acorn puffballs sit on a saucer-like disc of mycelium and sand grains, helping to keep them upright for more effective spore dispersal. Also spheroidal, earthballs differ from puffballs by splitting open to expose their powdery spore mass rather than having a central pore. Many earthballs are important mycorrhizal partners with plants. *Pisolithus* species are mycorrhizal and some earn their name, dyeballs, from the dyes that may be extracted from them.

It took me about eight years of going out into the field, returning with my collections and describing them, examining their spores and other microscopic characters often necessary for identification of species, before I could recognise that I had a new species. It can take years of examination and research before we can be certain that a new species has been found and can be formally described. It gives me a thrill when I find something that I am sure is a new species, yet to be described in the world literature. The first new species David and I found was in 2003 in Karte Conservation Park, a small park of sand dunes and low scrub about 240 kilometres east of Adelaide. It was a small, cream, stalked truffle-like fungus with a flattish oval cap. It was full of beautiful spherical spores ornamented with spiral whorls, earning it the species name of *turbinispora*. Its formal name is *Oudemansiella turbinispora*.

Another new species is a small, fan-shaped gilled fungus with a white furry top and white gills radiating from its attachment to rotten bark. I, David and Helen Vonow (Collections Manager at the State Herbarium of South Australia) found it in 2010 in Flinders Chase, growing on the underside of bark fallen from sugar gum *Eucalyptus cladocalyx* after a bushfire. The faintly pink tinge of its gills told me that it was a species of *Entoloma*, fungi whose spores are angled, looking rather like microscopic holly leaves. Small stemless Entolomas are rare and we were excited to find a new one. We published its description in 2016 and called it *Entoloma ravinense* after its location in the Ravine des Casoars in Flinders Chase. Since then, we have found it at only one other site in the park, and fear it may have been wiped out – its known locations were devastated by a severe bushfire in 2019/2020. We hope that it may occur elsewhere. A possible sighting in Victoria is promising but yet to be confirmed.

One strange fungus we found on Kangaroo Island was certainly a new species and did not belong in any known genus. It forms black rounded lumps 15–40 millimetres in diameter and usually grows in sandy soil in slight ditches on the edges of tracks. When cut in half, it resembles a tiny cavern, with chambers separated by stalactites and stalagmites and capped with a waxy black lid. We gave it the scientific name *Antrelloides atroceracea*: the first name (the genus name) means in the form of a small cave, and the second name (the species name) means black and waxy. We have found more than 20 new species, most of which are still to be described.

I enjoy going out and exploring fungi on my own and with David but, inspired by the example of Fungimap – especially after its very successful inaugural conference in Denmark, Western Australia in 2001 – I decided to start the Adelaide Fungal Studies Group. This group was set up under the auspices of the Field Naturalists Society of South Australia. I organised forays and meetings during South Australia's fungal season. On each foray, the group gathers with a leader, and everyone's shoes and boots are sprayed with methylated spirits to counter the spread of the water mould *Phytophthora cinnamomi*, which causes devastating die-back of native vegetation. A recorder notes the fungi found, and a list is assembled and sent to members and local park rangers. Sometimes we record as many as 60 species on a foray, but much depends on rainfall and other conditions. The shared experience gives a sense of accomplishment and we can look back and reminisce about when we found new, weird and beautiful species.

The Adelaide Fungal Studies Group is thriving, and is one of many groups nation-wide whose members survey and map the fungi of Australia. Having many eyes during a foray increases the number of fungi recorded at a site. Since its start in 2001, the group has contributed thousands of records to Fungimap which have been incorporated into the Atlas of Living Australia. The group also contributes photographs of fungi found on its forays to the iNaturalist project, a network of biologists, naturalists and citizen scientists which aims to map and share observations of global biodiversity.

Gradually, fungi are becoming more widely recognised as essential performers in the natural world and valued for what they can give in the way of food, medicines and an ever-increasing variety of applications – even for building materials. Internationally, projects such as the State of the World's Plants and Fungi by the Royal Botanic Gardens, Kew, assess our current knowledge of the Earth's plant and fungal diversity, global threats, and policies to safeguard it. Fungi are included in the International Union for Conservation of Nature RED list of threatened species. Societies and communities around the world now foster interest in the fungal kingdom – in Australia, these include Fungimap and the Australasian Mycological Society. The ever-increasing number of citizen scientists contribute to our knowledge of species' distribution. However, the value of their observations can be recognised only if there are enough scientific experts who can verify identifications and conduct relevant research. Field observations provide the groundwork, but laboratory work and research are essential to improve our knowledge and understanding of this diverse and complex kingdom and give it the recognition it needs and deserves.

Acknowledgements

I should like to thank the State Herbarium of South Australia for its support for more than 20 years, Dr Tom May and all at Fungimap for their professionalism and camaraderie, and the members of the Adelaide Fungal Studies Group for their enthusiasm and friendship. I am grateful to the Wildlife Conservation Fund for grants which helped to fund eight years of collecting. I am indebted to the many Park Rangers for their help and interest. I should especially like to thank my husband, David, for his continuous and active support in my mycological retirement.

5

A leopard at the nursery door

Neil R. Jordan

S elf-help books and websites distil a wealth of useful tips and tricks to help new and expectant parents and carers cope with a huge range of kid-related scenarios. Although my partner and I devoured them all frantically and prepared as best we could, raising kids in a remote research camp in Africa still threw us a few curveballs. Practical tips on how to politely refuse the well-intentioned, but overenthusiastic, advances of a stranger in a supermarket were thoroughly covered in the books. However, what should you do when the stranger taking an interest in your child is a wild leopard? Similarly, we read with interest about tricks to remove stubborn stains from cloth nappies but completely missed the section on how to avoid soiling yourself as a pride of lions converges on your camp, while you frantically attempt to soothe a crying infant. In the end, we learnt by doing. From our rustic tent in Botswana's Okavango Delta, we had to find our own way as parents, and learnt much about ourselves and our place in the world. Spoiler alert: everyone survives, including the wildlife!

As a newlywed but field-seasoned couple, Krys and I moved to Botswana in 2011 to work with the Botswana Predator Conservation Trust. The Trust is a local non-government organisation that gathers and applies information on large carnivore ecology to assist human communities and large predators to coexist. Iconic and idolised – but difficult to live alongside in reality – large carnivores represent a considerable conservation challenge, and as such they are a magnet for conservationists. The Trust was started in the late 1980s by Tico McNutt, a then PhD student studying African wild dogs, and its remit had since expanded considerably. It now encompassed the entire large carnivore guild – African wild dogs, lions, leopards, cheetahs and spotted hyaenas – all of which were present in the Delta in healthy populations. They all visited our research camp frequently too, reminding us at a deep level of our place in the world. While work–life balance is important, it was often hard to separate our professional focus – developing tools and strategies for human–wildlife coexistence – from our personal life in camp, which was essentially the exact same thing, just with higher personal stakes.

Tucked into the riparian trees alongside a remnant floodplain, about a two-hour drive from the closest town, our research camp was completely unfenced and accessible to our study animals and other charismatic megafauna. They had been coming here long before us and weren't going to let a few tents or a bunch of scruffy researchers interrupt their daily routines. It wasn't uncommon for a leopard to walk past the dining table as we ate our evening meal, nor for a thirsty lion or two to drink the runoff from our evening shower, with nothing between us save for a privacy screen of thinly woven reed and the confidence of youth. At night, our slumber was often disturbed by the gunfire cracks of snapping branches as groups of elephants browsed around our tent, brushing past the canvas walls, their gurgling bellies a noisy reminder of their near-constant need to eat. A honey badger even burnt down our kitchen one night, turning over the gas-powered fridge and lighting up the reed walls and canvas roof, and so while their fearless and troublesome reputation is not always deserved – I've seen them cowed by the mere shake of a broom towards their general vicinity – there's certainly no smoke without fire.

Camp was a magical place, where the term 'nature red in tooth and claw' often played out before us. It was also a constant struggle to keep it functional, and certainly wasn't the kind of place where your average parenting guide is super handy or practical. However, having learnt from necessity and researchers who came before us, we acquired the skills of self-reliance that are vital in remote off-grid living – indeed, Krys is now a seasoned Land Rover mechanic, solar electrician, and broom-wielding honey badger repeller. We soon found ourselves upping the stakes, adding a little one and then a second to the field team. Some at 'home' questioned why, struggling to conceive how we could even contemplate raising a family in the Delta with truckloads of 'hostile' megafauna around. Of course, generations have raised and continue to raise families in far more challenging situations. Indeed, we might reasonably claim that early humans 'grew up' under the same conditions, and many people continue to live out the full extent of their lives there without a second thought. We were well aware of our privilege. We were in a position to choose this raw and basic life, and could stop at any moment we chose. However, we weren't willing to give up on the experiences, opportunities and life-affirming struggles that this work and lifestyle offered – to be honest, I don't think we ever seriously considered not doing it. No height restrictions were in place on this rollercoaster, so why not bring the kids along for the ride?

Parenthood brings the unexpected. For us, a major change was the re-assessment and rekindling of our connection with our own place in the world, and a growing appreciation of the benefits that come with introducing children to it, allowing them to connect in ways we had not. As a child and teenager I spent long days in the woods of the UK, but obviously neither I nor my parents were ever afraid that I'd be carried off by a large carnivore. As long as I was back by dark my parents didn't seem to mind where I went. It was liberating, and to their credit they embraced, if not quite understood, my natural eccentricities. This acceptance included tolerating the abandoned bird nests and discarded eggshells crowding my windowsill, and the skulls and cadavers of various British mammals accumulating in shallow graves in the backyard, which I'd dig up months later to reconstruct their anatomy. These experiences were formative for me, but such freedom to explore was not something I'd be able to pass on to my own children in the wilds of Botswana. Indeed, to our horror we realised that we were, out of necessity, going to resemble helicopter parents, buzzing over our children 24/7 to keep them safe from harm.

The camp stretched out among the trees. Thin sandy walking trails connected the central fireplace, kitchen, dining and office areas to the satellite tents and workshop. Our day began with being woken by the pre-dawn chorus of noisy francolins, and we'd be drawn along one of the sandy paths in the early dawn light by the promise of strong coffee. In our pre-parenting days, we would interpret the signs and tracks of the species that had passed along the patch through the night. On some mornings, the dog-like tracks of spotted hyaenas would mingle with the callused saucers of elephant pads, their relative positions betraying which animal came first. On most days the news from the paths added detail to our half-wakened memories of the night's passing traffic. Like a commuter casually reviewing the previous day's events in the broadsheets, my pre-parent self was merely interested in reading this nightly news, but with parenthood this digestion of the daily bulletin became a critical part of our routine. We routinely swept and checked the trails and tent surrounds for signs of black mambas and other venomous snakes. As they grew, the boys came to enjoy this early reading lesson, learning the different tracks and how they guided our day to come, both in the camp and later in field. As they grew older, they began to accompany us more frequently during our field research, expanding their horizons while prompting us to reflect on our own.

For all of these growing changes, our world also shrank in many ways. The need for self-sufficiency required us to focus on somewhat menial tasks, using precious time that could have been spent on research. We also had to compromise aspects of ourselves and our parenting that didn't sit well with us as ecologists. Some of the local spiders and most of the resident snakes were not the best playmates for small grabby hands and curious minds, and so we reluctantly adopted the general approach that all 'creepy crawlies' were 'bad'. Although this premise was a compromise we felt we had to make to keep our kids safe, it was a complete antithesis for us, and we were concerned that we'd be raising a couple of arachnophobes. It was uncomfortable to have our tiny toddler make the regular walk from our tent to the main area while wagging his chubby finger and repeating his 'Spider. Baaad!' mantra every time we encountered a harmless invertebrate. He even gave a few sticks and twigs a comically wide berth. However, as the boys and their understanding of the world grew, their invertebrate education also become more nuanced, and I'm very relieved to say that they no longer make wide arcs around harmless ants and beetles. Quite the contrary, in fact. They developed a razor-sharp awareness of and interest in the fascinating denizens of leaf litter, and retain an ability to spot them almost everywhere. Our youngest son, in particular, has a growing love of bugs and other invertebrates, which has (to our great relief) been unfettered by the safety shackles of tactical misinformation we constrained him with early on. Perhaps the entomological forbidden fruit even inspired his interest – now, what a welcomed teenage rebellion that would be! We can but hope.

Along with these compromises, we had many moments of joy from this modern form of earthy subsistence. The floodplain was the boys' television, with an engaging and ever-changing cast. Sincere requests to 'Wilbetee. Deck. Go!?' communicated my young son's constant desire to go and look for wildebeest from the deck on the floodplain, and is a heart-warming memory for us all. They didn't need pets to learn the reality of death either. Death, and the life it sustains, was a major part of their day. They saw lions feast on rotting elephant, leopards steal the lunch of hyaenas, and African wild dogs tear 'Bambi' (in this case impala) apart against our toilet wall. They looked on with interest and inquiring minds. Even constrained to the camp, which they were for most of the time, there was so much for them to do and see. Their cultural exposure to researchers and staff from Botswana and elsewhere taught them much about the world. They developed and still feel a sense of home in nature, particularly in Botswana.

What they lacked in friends and pets they made up for with wild animals – from which they always kept a respectful distance. Vervet monkeys frequented the camp almost daily. Many a happy hour was spent by the boys lying on their backs on the deck, mesmerised by the gently swaying and ever-changing canopy overhead as birds perched and alighted, and vervets cavorted in the branches of lofty acacia. The vervets offered much more than entertainment, though – they became an important part of our security detail as their guttural and referential alarm calls warned us of danger nearby. Similarly, the baboons that roosted in the trees above the camp were a useful early-warning system, alerting us to the approach of leopards and lions in the night. However, as novel and earthy as our nursery monitors were, they weren't fail-safe, so we added an electronic baby monitor to our primate-based alarm system. The monitor was mostly used when we were in the main area of camp having dinner, listening for disturbances as the kids slept back in the tent. The monitor hung around my neck with the volume and sensitivity increased so high that in the early days we'd frequently re-tread the sandy path to check on the source of twigs being broken, or small branches brushing the canvas in a light breeze. Like parents everywhere, we became less and less attentive as the boys grew, and more confident and assured in our daily tasks and routine.

One evening, the lights on the monitor flashed to full illumination. At the same time, through the speaker, I heard a loud bump and brief but heavy patter on the tent's deck. The sound immediately caught my attention, so I grabbed my torch and headed quickly down the trail. The bump sounded too loud to be a genet – a small cat-like creature that could be tampering with the toiletries in our outdoor shower – and all the primates, apparently including our arboreal security firm, were asleep. While the possibility of a leopard had crossed my mind and contributed to my fast approach, I was still surprised as my torch picked up its lean figure on the deck, looking through the tent's meshy sections over the boys' beds towards my torch. Almost immediately it leapt from the deck away from me and melted silently into the bushes as if it had never been there, but the morning news showed its path clearly.

Dusty prints betrayed how the leopard had stood looking into the wide and flimsy shade-cloth door, which was zipped shut on either side and weighed down only by a narrow seam stuffed with sand like a draft excluder. Leopard-proof this door was not! This incident was by far the most alarming we'd experienced, and I walked into the tent to sit with the boys and catch my breath. They didn't stir, and continued sleeping soundly just 2 metres from where the inquisitive young leopard had stood watching them. Until this point, we had been comforted by the idea that predators saw the canvas tent as an impenetrable barrier and reassured ourselves that they wouldn't hop onto the deck to investigate through the screen. We felt that we knew our nocturnal neighbours, their routes and routines, but this leopard was clearly a newcomer – far more exploratory and arboreal than the old resident. The incident rattled me considerably, and I remained in the tent as the boys slept for the next few nights. Nevertheless, while it was one of those sliding-door moments that could have instigated a change in attitude about the place and our choices – and potentially far worse, of course – we soon grew accustomed to this new leopard's ways, slightly modified our bedtime routine and settled back into the rhythm of life. I can confirm, however, that having a leopard at the nursery door did much to underline our place in the world.

Humans have become accustomed to being perched atop the food chain. It is profoundly humbling and incredibly exhilarating to be knocked off that pedestal on occasion. While the oft-repeated phrase that 'children should know their place' is generally meant to encourage them to comply with society's expectations, I prefer to see this advice from an ecological perspective. It has been a privilege to immerse ourselves and our children in nature in this way. What purer way is there to find and feel your place in the world than by stepping back inside the food chain within which our species evolved? Like all parents, we make many mistakes along the way and question most of what we do, but with this and similar experiences etched onto their memories, I am confident of one thing: from an ecological perspective at least, our children know their place.

Acknowledgements

Thank you to the Botswana Government and local communities and community members – particularly from Sankuyo, Shorobe, Shukamukwa, Mababe and Khwai – for permission to work in this area. I am particularly grateful to Dr Tico McNutt and Lesley Boggs for entrusting us with the opportunity and support to become a part of this wild place and share in its wonders. Thank you to Helen Waudby, for inviting this chapter, editing it so well and giving me reason to relive this. Thank you too to Krys, Arun and Luka for allowing me to write about this and about them, and to the wildlife for reminding us of what we are and the responsibility that comes with that.

6

If the gators don't get you, the bugs will

Laura Kojima

When the COVID-19 pandemic hit the world in 2020, I was living alone in the south-eastern US, across the country from my family and loved ones. Despite the chaos and stress caused by the pandemic, I had to keep pushing through life as safely and normally as possible. Granted, nothing about the work I did was 'normal'. In February of 2020, three weeks before the world shut down, I had moved to South Carolina to begin collecting data for a Master's research project, which looked at how a contaminant called mercury could affect wild American alligators. You tend to hear about the risks of mercury in relation to eating tuna or other types of fish, because eating too much food with mercury in it can be bad for your brain, negatively affecting vision, muscle strength and overall movement. Animals like fish and alligators, which live in the water, are more likely than terrestrial animals to be exposed to mercury throughout their life cycle. Consequently, they often have high levels of it in their bloodstream, which can negatively affect alligators, but also the people that hunt and eat them (alligator meat is routinely consumed by some communities in parts of the US). My research focused on examining the link between mercury in the environment, in the bloodstream of alligators and in the humans who eat them.

The chaos of the pandemic created delays and much uncertainty, setting the project back timewise, but at the end of June 2020 we got the green light. As you would expect, capturing alligators is a group effort. Specifically, a group of four adults is needed to remove these large and dangerous animals from traps. I and another person would set and bait the traps earlier in the day and check them in the evening, when alligators are most active and likely to be trapped. The traps were a wooden contraption that, when activated, would noose the alligator using a wired snare attached to a rope that was tied to a tree, giving the alligators freedom to swim in the water but not to leave the capture site. Our capture group tended to be the same people each time; it was always myself and my two supervisors, Ben and Tracey, with one extra person who was usually another graduate student in our laboratory. Everyone in our group had been strictly quarantining and our field work was outside, so we weren't especially worried about the risk of COVID-19 exposure during trapping. Despite my only experience with alligators being holding a baby when taking a swamp tour in Louisiana, I never worried about being injured in the field. I trusted that I was in experienced hands with my group and that I would achieve their level of expertise soon enough, by being 'on the job'.

A week of setting traps and collecting data (like blood samples and body measurements) on the captured alligators proved to me that handling these animals is not as theatrical as the *Crocodile Hunter* may lead you to believe – but it's no activity for the faint-hearted either. Alligators are strong animals and we all made sure to treat them with respect. The first week made me feel like I could handle anything: driving a boat in rainy conditions, setting traps that required me to handle raw chicken, and looking for red glowing eyes in the middle of the night when my traps were nowhere to be found. Often, trapped alligators would take their anger out on the trap and destroy the wooden panels, leaving us to rely on a GPS coordinate to locate the contraption. Soon, it came time to move into the next phase of the study – attaching GPS transmitters to 13 adult male alligators. We wanted the data from these units so that we could look at whether variations in alligator movements may be based on their mercury blood concentrations.

Transmitter placement is a surgical procedure, carefully completed either on the airboat that we used to access the traps or on land near the trap. Alligators are not anaesthetised, like some other animal species are when receiving a surgical procedure, because it is difficult to calculate the level of anaesthesia needed for alligators accurately and they are at increased risk of drowning when anaesthetised. So, we had to rely on our strength to restrain the alligator while we attached the transmitter unit. A visiting biologist, Thomas, joined us for two nights of capture to help guide our team on attaching the transmitters. On the last night of Thomas' visit we went out to check the traps but caught nothing. It was a humid night in July, still surprisingly hot despite nearing midnight. We hunkered down in our airboat near the entrance to the lake, giving the alligators some more time to enter the traps. That's when we saw 2.5 metre adult swimming near our boat – excellent, it was the perfect size to receive a transmitter. Instead of cranking the boat's engine (which would scare the alligator off) Tracey and I grabbed paddles, guiding the boat to the alligator while Thomas stood on the stern with a pole snare in hand (a type of tool used to lasso alligators), preparing to capture the large alligator. Frustratingly, it spotted us and sank quietly underwater. We felt defeated, but then he popped right back up. This time we were not going down without a fight. We used one last trick we'd kept up our sleeve and searched for 'baby gator distress call' on YouTube. Baby alligators make a very distinct sound to let their mother know when they are in distress; this same noise can get the attention of other nearby alligators, who will come to investigate or help. Thankfully, our target alligator was a good Samaritan and turned in our direction, following the sound. With a few pushes of the paddle, we were able to get Thomas close enough that he was able to snare the alligator. The alligator only put up a little bit of a fight, doing the characteristic death roll of large crocodilians (a group that includes alligators and caimans, true crocodiles, gharials and false gharials) to rip the wire snare from his neck. However, once he settled we were able to get him safely on the boat where we secured his mouth and restrained his body. We confirmed that he was a male, to our relief. Male alligators move more than females, which provides us with more information for the overall population, so we were targeting the boys. Even if he ended up being our only capture of the night, I was grateful to have another 'transmittered' alligator on the books. The night could only go smoothly from there, right?

Thomas and Tracey restrained the alligator while I gathered all my supplies for the transmitter procedure and placed them in the order that I would need them. I numbed the area where the alligator would be getting the incisions for the transmitter, and Ben began the first part of the placement process. Thomas guided us through the procedures from his seat atop the restrained gator. I focused on maintaining my concentration, and did not expect what happened next. A small bug, likely attracted to the light emitted from our crew's headlamps and the reflections off the six earrings on my right ear, flew straight into my ear. Far enough into my ear that it couldn't get out!

I could not believe it. The timing was terrible. Who on Earth gets a bug in their ear right in the middle of placing a transmitter on an alligator?! I immediately stopped what I was doing and tried a few things to get the bug out of my ear. The fluttering of the insect's wings was an (unpleasant) orchestra in my head. I shone my headlamp into my ear, hoping to somehow attract the insect out of the canal, but without success. I put tape on the end of a pencil, sticky side up, and gently poked it into my ear to try and catch the tiny winged terror, but that didn't work either. I avoided asking for help for a few minutes, not wanting to interrupt the transmitter attachment process. Eventually, Ben noticed that I was distracted, and I confessed to what was bugging me. He offered to try to get it out with our tweezers and sterilising solution. However, everything I had read during my hurried online searches categorically advised against letting someone who isn't a medical professional mess with your ear. I declined Ben's help, worried about potential damage to my eardrum.

Panic began to set in. The bug was clearly feeling the same way, because it wouldn't stop buzzing. I couldn't believe the timing! Not just because were tagging an alligator, but also because we were in the middle of a pandemic ... Would I really have to put myself and others at risk of a COVID-19 infection by going to a hospital to get a bug out of my ear? An hour passed; it was past midnight at this point and the bug was still going at it. Usually, the loud fan of the airboat bothered me when driving around the lake, but that night it was a welcome distraction from the sounds inside my ear. I texted my boyfriend, who was on the west coast. He reminded me that he too had a bug stuck in his ear during field work for his graduate degree. An ironic coincidence, considering that the internet said that the phenomenon is rare. He warned me to not let the bug die and to stop sticking things in my ear because if the bug broke apart then pieces could get stuck and cause an infection. A kind of numbness from exhaustion began to suppress my panic, but I was still worried. By 3:00 am we were done checking traps and processing animals. It was time for me to go back to the laboratory to prepare and store my samples. I couldn't believe that I had made it through the night with the incessant buzzing. The fluttering was loud and constant, slowly making me go insane. Thankfully I managed to finish my laboratory work. I drove home and was showered and ready to go to bed by 4:00 am.

I know that most people would not go to bed with a bug in their ear but, like I mentioned earlier, nothing about my life was normal. So yes, I put my earplugs in. I can't sleep without feeling them in my ears because I'm a light sleeper. The bug presumably went to sleep too, as I was not woken up by its mad fluttering. I slept soundly, but when I woke up at noon the fluttering recommenced. I knew that I needed to go to hospital and get this thing out of my ear. I gave myself an hour to see if maybe the bug would, by some miracle, vacate my ear. I did not want to go to the hospital during a pandemic for an issue, which on face value, seemed so mundane. An hour passed, the trapped insect kept up its fluttering, and I knew it was time to mask up and get medical attention.

I went to the hospital, a 10 minute drive from my house, wearing two masks, a baseball cap and a hoodie. I had my hood pulled so tight over my head that I looked like Kenny from 'South Park'. Information about COVID-19 and how it was transmitted was still very limited at that point in the pandemic, so to say that I was paranoid about getting an infection was an understatement. It wasn't even necessarily myself I was worried about. I was really concerned about getting the virus and maybe passing it on to one of my advisors or colleagues. In my head, hospital = hotspot. I felt dread. After I checked in, I sat in a corner away from everyone else in the waiting room. Fortunately, the room wasn't too full and I didn't have to wait long. The nurse took my blood pressure and temperature in the waiting room, likely an indicator of the limited space they had at the time.

Another nurse escorted me to a more private room where I was to have the bug removed. This room was in the hall adjacent to the waiting room but blocked by a door. The degree of separation made me feel a little more secure and safe, particularly as the rooms in this hall had some privacy in the form of thick curtains. A physician's assistant (PA) came in, and after asking some basic questions it was time to remove the bug. He was concerned that I may have messed with my ear so much that I may have pushed the bug back until it was sitting on my eardrum. We both cringed at that thought. After a failed initial attempt at trying to suck out the bug with a syringe, the PA then grabbed two syringes, some lidocaine (a local anaesthetic to paralyse the bug) and saline solution to wash it out.

After squirting each of the liquids in my ear, the bug was flushed out. Instant relief! The PA let me take a picture of the winged beast, one in which I flipped it off for all it had put me through in the last 24 hours. The PA laughed; I could tell this incident was a slightly more light-hearted one for him than those he'd been attending lately. It was definitely an interesting situation for me. A bug in my ear was the last thing I would expect to send me to the hospital when working with alligators. Interestingly, we'd had all the supplies the PA used (bar the saline solution) on the boat. The lidocaine was used for the transmitter placement procedure, and I had the same syringes. Thankfully, I've never had to deal with a bug getting stuck in my ear again. Sometimes the biggest challenges you anticipate aren't the ones you'll face; sometimes the biggest challenges you'll face weren't even anticipated. If I learnt anything from the incident, it is to expect the unexpected, remain calm, and don't stick things in your ear.

Acknowledgements

I would like to thank my Master's advisors Ben and Tracey for their guidance and perseverance during my project. I would also like to acknowledge members of my gator-grabbing crew, particularly Samantha and Thomas for all their help during this project. Lastly, I would like to acknowledge the Savannah River Ecology Laboratory, the Savannah River Site and the University of Georgia's Odum School of Ecology.

7

Raising rays

Leonardo Guida

'I managed to grab five pregnant females today! I've got some amazing footage, wanna see?'

This exclamation sounds strange coming from a 30-year-old, 6'2' male without some context ...

In the warm shallows of southern Australia, just as the peak of summer has passed and its residual warmth slowly and lazily welcomes autumn, southern fiddler rays ('fiddlers') give birth to their pups in the shelter of seagrass beds. Each mother births up to six miniature replicas of herself and their father, each only slightly bigger than an average adult human's palm. They're beautifully adorned, being patterned with dark brown stripes across their mustard yellow body, almost like an ornate tribal tattoo. Each year, females congregate in these shallow warm waters to give birth and mate, fulfilling their role as marine ecosystem mesopredators (an animal that occupies the middle layer of the ocean food web) by eating the small crustaceans and fish below them, while avoiding being eaten by the larger sharks above them.

***

'Leo, there's a shark documentary on TV, come quick!' my father would often say. We both loved documentaries and the National Geographic shows about sharks were my favourite. I was 10 years old and the idea that sharks were in trouble wasn't really mainstream knowledge yet. It was the mid 1990s and shark fishing didn't peak globally and in Australia until the early 2000s. The idea of eating shark fin soup at the local Chinese restaurant was a curiosity rather than a concern. Shark documentaries focused on the awesome power of sharks, their biology, the dangers sharks posed to *us*, and the mystique they hold in our collective consciousness. Unlike today, shark documentaries back then – at least to my recollection – rarely if ever had a conservation bent.

Leonardo da Vinci was and still is my idol. As a child, his profession seemed a logical choice for me to follow. I had the name, I could draw and I loved science … Why would I be anything else but a scientist? Fast forward 20 years, and there I am on a boat in sun-drenched Swan Bay, Victoria. I'm checking that my scuba gear is working before diving into the water to hand-catch fiddler rays in the hope that the portable ultrasound on board the boat will reveal the flickers of embryos inside the mums-to-be. My goal? Investigate how the stress of being caught in commercial fishing gear influenced pregnancy, the pups-to-be and mum's general physical health.

In 2023, it's estimated that in Australian waters one in eight species of shark or ray, a group of animals collectively known as Chondrichthyes or the cartilaginous fish, are threatened with extinction because of overfishing. Globally, one-third of all sharks and rays face extinction, making them the most at-risk group of animals in the ocean. Like us they breed slowly, live a long life, and give birth to few pups – put simply, we're fishing them faster than they can reproduce.

Fiddler rays are unique to Australian waters, and fortunately they're not a threatened species. From an experimental perspective, they're a dream to work with too – they're abundant and easily caught by hand, safe to handle, easy to transport, are relatively hardy and handle captivity well. They're a perfect model species to study the effects of stress on pregnancy. At its core, I wanted to understand to what degree we were underestimating the impact of fishing on shark and ray populations. Would stress cause offspring to be aborted? Are they born small and sickly? Would mum be in good shape after giving birth (post partum) or would she skip breeding the next season so that she can recover from the trauma?

In the summer of 2013, I jumped on a boat at Queenscliff boat ramp and headed off into the channels of Swan Bay in search of pregnant fiddler rays. Based on existing literature and local experience, we knew that, come February, pregnant females would enter the shallows and be corralled by the receding water into the deeper channels of the bay for refuge at low tide. Once we passed over the channel and anchored, we pulled on scuba gear. Within only seconds of plunging off the side of the boat and into the water, there they were! Big beautiful females, bathing in the warm summer waters, living their best lives. Well, at least until a large dark shadow loomed over them and, with a quick flick of an arm, snatched their tail and gently swam back to the boat with them. Fortunately, rays don't need to move to breathe thanks to holes called spiracles just behind their eyes, so being propped under my right arm with one hand on the tail and the other hand gently under the belly made for a relatively calm experience for the rays and I.

On reaching the boat, I gently passed them up to a volunteer and once we'd collected a maximum of five (the number of rays that the tank on board could safely hold) I'd climb aboard and take off my scuba gear. I waved the wand of the portable ultrasound over their back to see if I could detect the flicker of an embryo in either of their two uteri. Plenty of pregnant female fiddlers were about, the diving was fun, and everything went as smoothly as it could possibly go. It was summer, and I was getting paid to dive and play with rays in my beautiful ocean backyard. I was (still am) living my best life.

Just down the road from the boat ramp where we launched from, I'd set up a makeshift maternity ward at the Queenscliff Marine Research Station where we kept the females over the next couple of months, until they gave birth. Each female was carefully transported, immediately measured, weighed and named. Apparently, you're not supposed to name your test subjects, but these rays weren't just fish, they were mums-to-be. Naming them seemed natural, and so each was named after a family member, friend, wife or girlfriend of a mate. My nonna Grazia and my three-year-old niece Natalie were especially chuffed to have pregnant fiddlers named after them.

The maternity ward consisted of large 19,000 litre tanks that could each house 10 female rays with plenty of room to spare. One of the tanks was equipped with a giant paddle that pushed water past stationary nets, into which we packed a group of rays. This process simulated a trawl net being dragged by a boat. Once they were 'trawled' into nets, we exposed them to 30 minutes of air to simulate conditions when fishers sort their catch on deck. Another group of females were the controls to compare against, being kept in a separate tank and not exposed to any treatment.

Over the next three months, each female was cared for. Twice a week, I fed them juicy pilchards and prawns, and gave them regular check-ups that included ultrasounds, blood sampling and weighing. In April–May the cutest little pups were born. Each mum gave birth to an average of two pups, one mum had five! The pups were measured and weighed, and we sampled a tiny drop of their blood.

I really felt a personal connection with the fiddlers. I noticed all the little details and quirks in their behaviour. When it was feeding time and I lifted the shade off the tank, they all swam directly towards me, likely having learnt that I was going to feed them (yes, sharks and rays can learn). Interestingly, the trawled females seemed to eat less, and I couldn't help but recall how I tended to eat less when stressed over long periods in the lead-up to high school and university exams. Often, at the end of a long day, as the golden sunset spread across the bay, I'd take a deep breath and watch them gently swim about their tanks or simply lie still. They were breathtaking and I loved them.

I was not prepared for what came next. It was an early autumn evening; the sun was half set and I had just finished cleaning the tanks after a long 12 hour day. I wiped the sweat from my brow, closed the tank valve, turned on the pump, and on my way to the car grabbed a meat pie from the kitchen for the drive home. When I returned early the next morning I was greeted by a devastating scene. I lifted the shade on the tank to say hello to the rays – and found seven dead mothers lying next to the drain in the tank. I hadn't shut the valve properly. Although water was being pumped into the tank the whole time, it drained faster than it filled. Only three mothers survived, huddling in a divot in the tank floor that allowed water to pool just above their eyes and was barely deep enough for them to pump water through their spiracles (remember, the holes just behind the eye) and over their gills. My mind raced and my body froze, in a panic. 'What do I do, what do I do, what do I do?'

I reported the event to my supervisor, describing every little detail, trying to piece together how it happened and, more importantly, how I could stop it happening ever again. I fully expected my supervisor to berate me. Instead, I was comforted with a reassuring smile, which said, 'You are human, Leo. Mistakes happen. Here's how we make sure it never happens again, not just for you but for anyone in the future'. We installed fail-safes in the valves, made waterproof checklists as reminders, documented the new protocol for future projects, and reviewed the incident and proposed changes with the university's animal ethics committee.

The lessons learnt in this experience extended beyond simple tank mechanics and tweaking of laboratory protocols. I was deeply connected to the rays and had always felt responsible for them, but their deaths really drove home my sense of obligation to protect marine wildlife. Fish are not 'just fish'; while they may not be the archetypal cute-and-cuddly animal we humans tend to gravitate towards, they are intelligent, graceful and awe-inspiring.

The experiment was ultimately a success. Like humans, severe stress during pregnancy resulted in smaller offspring and the mothers' post partum body condition was poorer. On a physiological level, stressed mums had chronically suppressed reproductive hormones, and their offspring showed elevated immune responses. I was also able to show for the first time that even though sharks and rays can survive capture, it can compromise their reproductive health and the survival of their offspring. This work demonstrated how fishing impacts these animals even when they are released.

***

It'd been around a year since I first glimpsed a beautiful pregnant fiddler ray underwater. I was in the final stages of writing up my results for publication and was excited to put *my* baby out into the world and showcase not only how resilient these rays are, but also the story they have to tell about how we fish and the pressure it places on shark and ray populations. 'I'm about to publish my best work!' I thought, and just as I was about to submit my paper, my dad attempted suicide. My world crashed. Hard. Dad was my biggest support in every sense and the roles had been suddenly reversed, not just for me but for my two brothers. My thoughts spiralled. 'Can I finish my PhD? Dad needs financial, emotional and physical support. I need to get a "proper" job. But my PhD! Should I be selfish and leave the heavy lifting to my brothers? I've worked so hard to get here!' Anger and sadness boiled over. 'WHY!?'

Full-time work was my only option. My PhD had to take a back seat. I remember being simply emotionally and mentally exhausted with life and said to my supervisor, point blank, 'What is the bare minimum I must do? I want out in the quickest and easiest way. I'm tired and I'm over it.' For quite a while I felt like I was wading chest-high through a lake of mud as thick and dense as molasses. Each day after my job working at a company that made medical devices, I swung past my university office and chipped away at my thesis, a few hours at time. For a little over a year, I persisted slowly but consistently, and in that time my paper about pregnant fiddler rays was finally published. I was stoked and relieved to say the least, because in that moment of receiving the email of acceptance for publication, the penny dropped. 'I can do this.'

The night when I got home, after receiving the news about my paper, I sat outside and cracked open a beer, staring at the stars. In the silence and enveloped by a warm summer night's breeze, I reflected on my work. The rays had taught me to see beauty where many would not, and they'd reinforced the importance of honesty and compassion when, because of me, some had died. My parents had raised me to follow my passions, believe in myself and, no matter what, always get back up after a setback. My supervisors had helped me refine the invaluable skills of critical thinking, communication and project management, but at my lowest points also reassured me that it was okay to stuff up.

With my path clearer and the end to my PhD in sight, I felt a renewed sense of passion and energy. I had wanted to do a follow-up study that looked at how pups from stressed mums behaved. Did they swim slower? Did they eat and grow as fast as their unstressed counterparts? Did they respond to predators more slowly? I had the whole experiment designed, approved, and was ready to go. Next season arrived, and I jumped in the water. Same time. Same place. No fiddler rays. The season before, I'd catch 30 pregnant females in any given week, but no matter my efforts this time around, I couldn't catch a single ray.

That season the seagrass had boomed and was as thick as a forest. The water temperature had peaked earlier in the summer and been warmer for longer than normal. Seagrass forms the base of the marine food web and provides crucial shelter for rays and their pups. With plentiful habitat and food, and warm water to support the growth of pups, we assumed that the fiddler rays had shifted their reproductive schedule to complete it earlier. Fiddler rays are capable of embryonic diapause – like kangaroos, they can control their reproduction when conditions are just right. So rather than wait until a little later in the year, they took advantage of increased productivity in the environment. I'd probably missed them by just a month or two. I had to halt the project, shelved the idea and focused on finalising my thesis.

Sometimes, you have to accept that you can't do it all. Life goes on around you and waits for nobody. I've found that accepting this reality is liberating and empowering. Instead of exhausting myself trying to do too many things and seemingly getting nowhere, I can *really* choose where my energy and passion go – be it other people, a body of work, or a hobby. I use that energy to dive deeper and that's where I've found rewards and fulfillment. You control your fate.

***

The end of my PhD was the beginning of my biggest and best adventure yet. Dad was with me at my graduation, and he was as proud as punch! We'd had a huge journey, with plenty of ups and downs, and now we talk about each other's mental health all the time. Our relationship and our lives have never been better.

The experiment I shelved was picked up by the next PhD student in the laboratory and they really made it their own. Their work demonstrated that pups from stressed mums were slower to bury themselves in sand and less likely to evade a potential predator in time. I was so proud of and happy for the student – I had seen an idea born, grow and become something all its own, thanks to the creativity, intellect and passion of another.

I'm now based at the Australian Marine Conservation Society, leading its shark conservation work. I still pinch myself every day at my good fortune. What I love most about my work is that whether it's the dry process of dealing with administration or the vibrant energy of being in a classroom, I get to 'geek out' about just how amazing and important sharks are for the health of our oceans, and by extension our own health. Certainly a career in conservation has many downs, but the ups make it worthwhile. Resilience, perseverance and perspective – those are all skills I refined throughout my PhD and continue to use in life and my job.

To those embarking on your scientific journey, mistakes will be made, failures are certain, and life will march on with or without you. Should you choose to undertake a PhD, know that your biggest success will be *you* and all that *you* will become.

Acknowledgements

I'm fond of the African proverb 'It takes a village to raise a child'. So, I'd like to thank my family and friends, colleagues past and present, and my mentors for shaping me into the person I am today and for the wonderful journey I've been fortunate to have. Thanks also to all the fish out there who made every dive unforgettable!

8

In pursuit of pollinators

Manu E. Saunders

As a researcher, it's wise to evaluate your choices regularly, whether they be career opportunities, collaborations or research methods. Moments of self-reflection – those contemplative 'Why am I doing this?' type of moments – are necessary and should be indulged.

I had many such moments during my PhD in ecology, which was a new career path after years of working in retail, corporate communications and warehousing. Having grown up near the ocean, I was deeply attached to the humidity and lush productivity of Australia's east coast. Then, I crossed the Great Dividing Range* and travelled south to start my research career in ecological science. My study system was the hot dry mallee of north-western Victoria. This region was originally a rich mosaic of woodlands, lakes and riverine forests that supported the traditional lands of many Aboriginal peoples, including the Jarijari, Watiwati, Latjilatji, Barkindji, Muthi Muthi, Wembawemba and others. The word 'mallee' is thought to derive from a Wembawemba word 'mali' meaning 'water', in reference to the unique ability of the region's multi-stemmed eucalypts to hold water in their large lignotubers. Mallee is a type of tree, a type of woodland and a region – a truly cross-scale ecological term. My research questions required me to spend many days and weeks in the mallee, traipsing through dusty monocultures of almond plantations and scrubby woodlands, collecting pollinator insects. It is where I experienced many moments of self-reflection.

I was investigating how intensive almond plantation management affected local communities of native pollinating insects (bees, flies and

---

* A mountain range that runs roughly parallel with Australia's east coast and is one of the longest land-based mountain chains in the world.

wasps). The region has suffered years of degradation and human disturbance since Europeans moved in. Vast agricultural landscapes encroach on remnant patches of scrubby woodlands that shrink year by year. Some of the crops grown here on a vast scale, like almond, depend on pollination for yields, but the intensive management practices favoured by plantation managers don't provide enough habitat and food resources to support wild pollinator communities year-round. Instead, crop yields depend on expensive annual honeybee rentals for the brief period that almond plantations erupt into mass bloom. When I started my research, this management practice was standard across most of the almond industry as little was known about the native insects that could potentially do the job of pollination for free. I had designed this project to combine my interests in conservation and agriculture and my love of wide-open spaces. I recognised the need for more knowledge on how to improve agricultural practices to support biodiversity, and saw an opportunity to address these problems in a unique ecosystem. 'Why am I doing this?' Like most ecologists, I wanted to do my bit to make the world a better place.

The Victorian mallee region is an enigma, a landscape of change. The rolling dunes found here are remnants of ancient sands from Australia's arid centre and are usually aligned east–west, moulded by centuries of strong winds roaring inland off the Southern Ocean. I found this corner of the mallee torn between two histories – it is an ancient landscape perpetually trying to outlive modern human disturbances. The flotsam and jetsam of human settlement floated about the mallee but was not part of it. Dilapidated plasterboard shacks were propped up next to immaculate Mediterranean-style mansions. All looked out of time and place next to the pockets of wild mallee scrub around them.

The mallee is a surreal and beautiful wilderness, where the closeness of the woodland around you can make you feel like the only person on Earth. Of course, that is until your reverie is interrupted by the sight of a rusted relic of past European invasions – abandoned bulldozer chains, rusted water-tanks and antiquated car parts. I felt a desolate loneliness in many of the mallee reserves adjacent to plantations when I saw these left-behind items, stark reminders of the abuse that the mallee has suffered. I rarely saw native birds and mammals, although I spotted plenty of feral pig prints in the sand, alongside four-wheel drive vehicle tracks, discarded beer bottles and bullet casings. As a woman working alone in a remote region with its fair share of aggravated crimes, I was always on edge and wary of every strange sound, especially the hum of approaching cars in the middle of an isolated woodland. 'Why am I doing this?'

One of my main sites was Annuello Flora and Fauna Reserve, a mysterious woodland that bordered some of my plantation sites. Little had been written about the reserve's history and environmental significance, although most locals could share oral histories and anecdotes of the area. It was once cleared, but the mallee fought back and after years of regrowth it was annexed to Murray-Sunset National Park, to which it was connected by a corridor of remnant vegetation. This foresight created one of the biggest planned corridors in Australia and it was a relatively stable home for many threatened species. One afternoon I drove around a bend in a bleak almond plantation, right into its southern corner where it hugged Annuello's core and saw a pair of malleefowl strolling calmly along the fence line. It was the first time I had seen these large noble birds in the flesh. These moments made my struggles feel worthwhile.

Like most of the mallee, Annuello had a dubious history of use and abuse before it gained protection. Clearing began in the early 1900s, when the 'monotonous' mallee scrub was seen as an inconvenient wilderness that needed to be conquered in the European war against nature. As returned soldiers and other new arrivals tried to scratch out a living from these depauperate lands, acres of mallee forest were burned and uprooted, tenacious roots ripped from the soil they'd clung to for centuries. The ecological and spiritual value of this habitat was realised many decades later, perhaps too late for many species, and the young mallee woodlands I walked through still bore the scars of this battle.

Early one winter morning I was walking deep in the interior of an almond plantation, setting out insect traps. These plantations were thousands of hectares in size, with neatly arranged rows of identical almond trees and meticulously cleared dusty ground. In winter, when the crooked arms of deciduous almond trees were bare of leaves, not a skerrick of green could be seen. This morning was eerily silent, with fog hanging low over the trees. Visibility was only a few metres. I'm a skilled navigator, but as I walked down my transect line I realised how easy it would be to become disorientated and lost here. Every row, every almond tree, looked identical. Take the wrong turn in this fog, and I could be walking in circles. I hurried along, stopping at measured intervals to put my insect traps out. As I placed the last trap down, I heard an unexpected noise. It was mechanical and getting nearer. I walked faster back down the row towards my car. The harsh mechanical noise drew closer. For a second, the fog swirled and I saw it – a mechanical pruner barrelling towards me, its long arms slicing and dicing as it went. I knew the driver hadn't seen me. I ran to get out of the way, my heart pounding in my chest.

'Why am I doing this?' I cursed the plantation managers and my career choices. This incident wasn't the first time my field work had been interrupted. Every trip, I notified the managers of where I would be working, signed in and out of the site office and followed their safety protocols. Managers were supposed to notify any on-ground workers of my whereabouts, but it never seemed to happen. Some of the workers I encountered unexpectedly were friendly. A tractor driver once stopped to chat to me, curious about the coloured bowls he'd seen around the plantation. He had never heard of native bees and assumed all insects, except honey bees, were pests, so he was surprised to discover I was trying to encourage insects to hang around. Other encounters weren't so friendly. I've worked in orchard systems across multiple regions and have been wolf-whistled at, sprayed with fungicide, had my traps mowed over, been caught in conversations about chemtrails and government conspiracy theories, and had passive aggressive comments about 'greenies' thrown at me. Thankfully, these encounters are rare, but they had a disproportionate influence on my perspectives at times. I had chosen this career path because I always felt more comfortable communing with nature than with people, but I was beginning to realise that ecological research often has far more to do with people, and people problems, than the plants and animals I was studying.

It was lonely dusty work in the almond plantations. This loneliness always drove me back to the wide brown Murray River, where the dry mallee sand dunes gave way to grey clays and regal river red gum forests, thin strips of opulence in the midst of desert. Whenever I had a break in field work, I made a beeline for the nearest riverbank in the Hattah Lakes and Murray-Kulkyne network of nature reserves.

One day in Hattah Lakes National Park, I encountered one of the most magnificent trees I had ever seen, rooted about 6 metres from the water's edge. The trunk had a huge girth, signifying great maturity. However, its most striking feature was the errant branch (it could almost have been a secondary trunk) which sought hydration so fervently that it emerged from near the base of the red gum's trunk and curved in a perfect architectural arch, high enough for me to walk under, until the growing end was submersed in the waters of one of Hattah's many lakes. The entire top surface of the arch was thick with sprouted foliage, little saplings creating a curtain of elegant leaves leading to the water's edge. Many times, I found myself standing by this tree in awe, infusing my soul with its strength.

The Ramsar-listed Hattah Lakes teem with birds most of the year and provide critical holiday habitat for many migratory species. The lakes are anabranches of the Murray River, rebellious pools trying to gain independence from the river before destiny defeats them. The water shimmers in blue and silver, lapping softly at the eucalypt-lined shores – a strong contrast to the red, green and gold of the mallee dune country above. In my second year of field work I arrived in the winter of 2011, when the Millennium Drought was declared broken. The lakes had flooded the tortured red gums just outside the water's usual reach, cutting access along nearby tracks. The surplus moisture seeping through the ecosystem had drawn animals from far and wide, and I noticed immediately that the choir of birds had strengthened since my last visit. Regent parrots and mallee ringnecks twittered alongside corellas and pink cockatoos. Wandering scarlet robins and flocks of zebra finches darted through the scrub, teasing me with their presence before flittering away too fast for me to follow them. For a girl from coastal Queensland, who grew up with seagulls, coucals and catbirds, I was wide-eyed with wonder. 'This is why am I doing this!'

Silence envelops the dune country in Hattah, but it is an expectant silence, as if something is about to happen. Out in the middle of Hattah's mallee dunes true silence doesn't last, as the wind scuttles through leaves and rattles dried hanging branches and seed pods. If I stood still in a clearing, I could hear it rushing at me across the landscape, like water surging through a barrier, carrying the whisper of mallee stories past and present. Hattah feels more 'alive' than Annuello. The greater plant diversity alone gives Hattah's landscape a vibrant glow that is lacking in Annuello's northern frontier. Stands of sedate cypress pine blend seamlessly with patches of unruly mallee eucalypts, all linked by patchy carpets of herbs, lichens, cryptogamic crusts or wildflower meadows. In Hattah, the air is laden with the promise of life, hidden in the interstices. As I drove through the park, emus dissolved out of the trees in twos and threes, strutting gracefully until surprise caused them to break into their more usual ungainly gait, feathered rumps bouncing from side-to-side as they ran. Red kangaroos browsed in lush mallee meadows that were cleared decades earlier to support the grazing of hard-hooved livestock, blending seamlessly with the red soil around them.

These moments reminded me why I was there. Whenever I felt like giving up after a failed trap, an unsafe encounter or a miserable day in the plantations, a trip to the river always boosted my strength.

My field work eventually came to an end. My data showed that wild pollinator species were around, but due to the lack of habitat and food in intensive almond plantations these pollinating insects were mostly restricted to mallee woodland areas around the plantations. Some pollinators ventured into plantations during the mass bloom period that lasts a couple of weeks in late winter, but they didn't travel very far from the woodland edges. Without drastic changes to management practices, it was unlikely they would provide much free pollination service to almond plantations.

Although I no longer need to visit the mallee regions for field work, the place has connected deeply with my soul and I take every opportunity to return. The most rewarding part of those field trips were the moments of reflection that reminded me why I was there. Research careers are exhausting, they can be disheartening and uncertain at times, and they can lead you down paths you never envisioned. But amid all the turmoil, connections with special natural places are often the muse that will remind you of your goals and purpose.

Acknowledgements

This story was made possible with the support and patience of my PhD supervisor Gary Luck, the team at Charles Sturt University's Institute for Land Water and Society (now Gulbali Institute), and my partner James Abell. Thanks also to the friends, family, colleagues and landholders who helped along my research journey, and my PhD co-supervisor Margie Mayfield who encouraged me to take a research leap into the unknown.

9

Of absences and Amazonia

David M. Watson

I consider any day I cross paths with a snake to be a good day but, in March 1997, in a remote tributary of the Amazon, I had a short encounter with a long snake that changed my life. Not one of those 'I never knew about this bizarre creature and now I'm devoting my entire career to learning everything about them' kind of revelations; it was an emphatic nudge that yielded new perspective on one of the questions that would define my science. I should also probably mention that I was naked at the time.

***

The scene played out in Iwokrama, a legendary expanse of northern Amazonia that ranks among the world's most diverse and least accessible places. I was part of an international team of ornithologists working with Guyanese students and Makushi tribespeople to conduct the first thorough survey of birdlife inhabiting this rainforest. I was halfway through my doctoral studies and trying to answer a deceptively simple question: why did some places support a greater variety of wildlife than others? I'd first tackled this topic for my Honours project in western Victoria – a land of bleating sheep, boom-and-bust wheat and a particular kind of woodland. Rather than eucalypts or acacias, the trees that defined the district were stately bulokes. Once-continuous woodlands had been cleared a century ago, and the remaining patches ranged from isolated postage stamp-size fragments to meandering roadside stands. Larger blocks didn't necessarily make better habitat, with some of the smallest patches being the only places where especially picky birds persisted.

However, when the counting was done and the number crunching complete, I considered the bigger picture and realised that my findings were irrelevant. Thinking back through all my 20 minute bird counts – 243 of them – plus my 30-odd camp sites tucked away from the road, I realised I'd only seen a handful of saplings. If the habitat I was studying wasn't regenerating, what did it matter whether larger or less-isolated bush blocks temporarily contained a few more birds? The patterns I'd so carefully enumerated were fleeting, transitional dynamics as these slow-growing trees adjusted to the imposition of a new management regime. Much older landscapes were needed to address this question; old enough for habitats to settle into new arrangements, for mobile animals to reshuffle and for less mobile things to either thrive or die out. So, I moved to the USA and selected the cloud forests of Mesoamerica as my new study system. I spent many months in the misty embrace of these mountaintop relics from the last ice age, establishing which species lived where and crafting a generalised explanation for the checkerboard distributions I was mapping.

My PhD advisor, Town Peterson, was an ecologist with extensive experience working in Mexico. He was one of two curators of ornithology at the Natural History Museum at the University of Kansas, so my office was in the museum. It was a whole new world for me, and what I initially presumed to be row upon row of map cabinets turned out to be specimen cases housing over 100,000 birds from all corners of the globe. Showing me around my new workplace, collections manager Mark Robbins explained that he led regular collecting trips, strategically filling gaps in the collection, but only people who were proficient at skinning birds were invited to join him. South America was a priority region for the collection – indeed, one of the main reasons for my crossing the Pacific was to fulfil a childhood dream of exploring South American forests and seeing their fabled fauna first-hand. So, I devoted myself to learning the fine art of specimen preparation and, a couple of years and a few hundred bird specimens later, received a tap on the shoulder – the time had come for my first scientific expedition!

Just getting to the site was a saga involving a repurposed mining truck that ploughed through potholes of such epic proportions that they inspired a documentary entitled 'The Worst Road in the World'. After driving all day and all night, we finally crossed the river and dropped off our gear. We shied away from the glaring sun and walked into the forest, beckoned by the shade and the promise of what it held. It was quiet, dark and utterly still – a completely different world from the bustling village of Kurupukari just a few paces away, on the edge of the forest. It took my eyes several moments to adjust. It was so dark on the forest floor that few plants grew there, lending it an open, almost manicured appearance. Then there was the smell – part spice rack, part freshly turned soil, part something else – just beyond my olfactory lexicon, approaching the mustiness of one of the less charismatic zoo exhibits.

Being there with Mark was something else. While I was slack-jawed and sniffing, he was looking up and smiling, cupping a hand behind one ear and tilting his head from side-to-side, checking off the bird species he could hear all the way up in the canopy. In the days ahead, he pointed them out to me, introducing them like old friends. The rhythmic pulse of scaled pigeons; the ethereal morning call of great tinamous, their tremulous cadence hanging in the air; the staccato of cream-coloured woodpeckers hacking at rotting palms for the succulent grubs within.

Our team included seven members of the indigenous Makushi people, the original custodians of these lands. This trip was my first time working with First Nations people on country, and each day brought fresh revelations. Within a few hours of choosing our first campsite, three men cleared an area of undergrowth, trimmed trees, notched poles and cut crossbeams to make a framework to support taut tarpaulins and a row of hammocks for sleeping quarters, as well as some benches and tables from the offcuts – all made then and there with their machetes. When we walked up an isolated mountain searching for white bellbirds, they made comfortable backpacks from palm fronds to carry water, discarding them on the forest floor when we returned and they'd served their purpose. When a few of my insect bites became swollen and tender, Milner, one of the Makushi men, told me to come and find him before I went to bed. He examined the bites, then applied small gobs of sticky sap, similar in texture to chewing gum, from a vine he'd harvested earlier. He then told me to see him first thing in the morning. The next day he prized the dried sap off the bites and squeezed the skin beneath – hard enough to expel the botfly larvae that had been starved of oxygen, the largest of which was the size of a cashew. He took great joy in showing me the rings of tiny black barbs around its body, used to cling tightly to the void it had been eating out of my flesh, chuckling as my face paled.

After a few days cutting trails and becoming acquainted with the mix of habitats around camp, we settled into a routine. Two people were on mist-net detail, walking the loop, extracting birds, leaves and the occasional beetle from the 20 nets, and bringing any notable birds back to camp where most were photographed and released. Another person was on collecting detail, walking trails far from camp with a tape recorder and 16-gauge shotgun to collect species for which no genetic samples existed. Everyone else stayed in camp and prepared specimens. As well as study skins, anatomical specimens were our priority, as our captures were often the first representatives of their species and critical from a scientific perspective. Once the nets were furled at dusk, everyone worked on specimen preparation, meticulously preserving the day's catch in readiness for the next. Rather than discard them, we saved the carcasses of any birds larger than a starling for meals, adding some variety to our curried beans and rice. Although I've eaten a variety of Amazonian birds – from toucans to macaws, ant-shrikes to oropendolas, fruitcrows to potoos – they all tasted like curry.

After dinner, when all the ornithologists were together, I reviewed the daily tally and read out a growing list of every species we'd detected. Every day, as well as updating the list with new additions, every species was scored and given each a tick or a cross for that day: detected or undetected. My PhD advisor was developing new ways to estimate how many species lived in an area using cumulative plots of detections to predict the total number of species present. Called 'species richness', this number encapsulates the diversity of a place, allowing different sites to be compared and changes tracked through time. He'd trialled the method with various datasets, including daily lists of state number plates spotted on his morning commute, to discover what equations yielded more reliable predictions of the true number. In a vast tropical forest with 500-plus bird species that were mostly brown and shy, this expedition would help refine the approach.

Our second camp was a whole day's journey upriver by dugout canoe. The view from the river was refreshingly different – you could see the undulating canopy, and finally catch sight of the flocks of swifts we'd heard since we arrived, careening high above the trees in search of flying insects. A large set of rapids required all of us to get out of the canoes and lighten the load so that the expert boat drivers could rev the outboards and power through. Watching the liquid nitrogen tank that contained so many unique tissue samples, a trove of data for generations of future scientists, rocking back and forth during an especially rough patch, all at the whim of a single unseen boulder, was heart-stopping.

Having the river close to camp was welcome. We spotted sunbittern and lyre-tailed nightjars most days; boat-billed heron and giant green ibis lurked amid the rocky meanders; loose skeins of blue and gold macaws flew along the river in the afternoon, their cries echoing along the fringing vegetation in their wake. As well as new birds and fresh fish, the river afforded convenient opportunities to bathe. Between the tropical heat and long days of specimen preparation, personal hygiene was a priority. The first time we waded right into the river, but our Makushi colleagues explained that this approach was unwise – various viper species used the tangled banks as prime hunting grounds and stingrays lurked on the sandy flats. It was far safer to pile into a canoe and paddle out to the middle of the river, where large granite boulders protruded from the café latte-coloured water. It was a case of choose your boulder, strip off and have a scrub. The women had the boulders on one side of the boat, the men had the other.

Every part of these bathing sessions was unnerving. There'd often be a big splash when we arrived at the rocks. Not only were they feeding stations for giant river otters, but we'd seen river turtles, sometimes more than a metre wide, hauled out to bask on the boulders. Big animals, big splashes. Black caiman also lived here. Formidable predators. People-eaters. Harold, one of the drivers and Georgetown fixers working for the National Academy, told us about a massive caiman with a mean streak that lived in a waterhole beside his base camp when he was a surveyor, putting the first lines on a blank map as parts of Guyana were originally charted. It was a favourite story and we'd all heard him tell it several times, replete with details about how she'd lie in wait. So, every time we heard a splash as we approached the rocks, one of the fellas would say,

'Hey Dave, she's waiting for you, man!' I knew they were fooling, but my pulse always quickened that little bit more. Stepping off my boulder and taking care to keep my balance, I was met with the unnerving sensation of dozens of small fish nibbling my skin – tetras! I don't know if they were getting anything or if they were just seeing freckles and investigating further, sometimes giving the slightest of exploratory nips.

So there I was, stark naked, senses heightened, balancing on a slippery rock in murky waist-deep water, reminding myself that the pinches were just happy little fish and that big splash I heard earlier was just a friendly turtle. I'd just decided that I was probably clean enough and could get out of the water, when something large bumped my leg. I looked down to see a waterlogged tree trunk just beneath the surface, something we'd occasionally see floating past. Looking closer, I was intrigued to see an interesting angular blotchiness to it as if a fungus or maybe slime mould was growing in a regular pattern almost like a mesh. I saw another diamond-shaped area, and then realised it was a repeating pattern of scales. By the time I recognised that the tree trunk was actually an anaconda, it had sunk out of view. I don't remember getting out of the river, but I do remember noting that the snake's muscular bulk was thicker than my thigh. Much thicker.

***

Reflecting on this moment, and the fact that a dozen of us had been bathing in this stretch of river for weeks, blissfully ignorant that we'd been sharing it with the world's largest snake, I thought back to all those crosses in the daily tally – the birds we knew were living around us but weren't encountered that day. As well as yielding reliable estimates of just how many different kinds of bird or bat, orchid or ant occur in an area, the crosses or zeroes in the surveys speak volumes about the organism itself. How much work you need to put in, how many times you need to look and find nothing before you can be confident that the object of your search is really not there. Detectability matters. Some species – kookaburras, cockatoos, currawongs – are easy to notice. Others are sneaky. Quail, cassowary, koala – you might not have seen one in your regular survey, but that doesn't mean they weren't there.

Once you grasp the concept of detectability, you can't help but see the world differently. The concept relates to climate change and food security. Fish might be moving to cooler waters, but they might also become easier to catch in warmer waters, so it's probably worth double-checking the numbers before announcing new quotas. Detectability also links to estimates of extinction – working out what species persist versus those that have already blinked out. It's no longer about what's obviously there, it's about what you might need to spend more time seeking out. What might be right under your nose, and you don't even know it was there.

Ecologists think a lot about methods to minimise bias. About what nets to use, which depths and seasons to survey, which correction factors to include in our statistical models. However, as well as enhancing objectivity, the concept of detectability embraces subjectivity. Different people see different things. My colleague Milner could see more with his naked eye than I could with my fancy 10× German binoculars. He spotted a backlit trio of blue-backed tanagers atop a 50 metre tree on the edge of Camp Two. The first tissue samples of that genus came from one of those birds. The rich diversity of life forms we strive to safeguard is just as valuable as the diversity in ways of seeing and thinking. Diversity matters, as does amplifying First Nations voices as we broker new ways to reconcile humanity and biodiversity.

Since that expedition, I've designed new ways of conducting ecological surveys, letting the results dictate when the estimates are reliable, rather than hoping that the allocated effort is sufficient. As well as optimising field work, these approaches have allowed very small and very large areas to be compared directly, clarifying what characteristics matter most in explaining diversity patterns. I've applied this approach retrospectively to long-duration recordings and geographically to vegetation transects. Along the way, I've listened to Wiradjuri and Dja Dja Wurrung Elders, learned from Badimaya and Nanda peoples the true meaning of caring for Country, and driven way off track in the Western Desert, clinging to the roof rack of a ute while being regaled with Martu dreaming stories describing how the landscape came to be. And although I've returned to Latin American forests many times, swimming in rivers, wading in wetlands and struggling through vine tangles, I've never seen another anaconda. I wonder how many have seen me.

10
Caving for spiders

Jessica Marsh

Finding yourself 20 metres underground, in an enclosed space, up close and personal with a very large long-legged spider is not everyone's cup of tea. For me it was the experience of a lifetime.

My adventures with caving started when I was young but were short-lived. I have memories of squeezing into a thick wetsuit and climbing down into a small dark hole in the ground with my dad, an avid caver. I grew up in Sheffield in northern England. Caves there were wet, cold and muddy. Years passed, and the next time I ventured into a cave, Australia was my home, my early caving expeditions were a distant memory, and I was searching for some large, rare and very elusive cave-dwelling spiders.

I am an arachnologist, working on spiders. Why spiders? I think the answer is simply that I have always been intrigued by them ... I was that kid who always had a snail, an insect or a spider in a jar. As is often the case with these things, the more I learnt about these little invertebrates (animals without backbones), the more I wanted to find out. It wasn't long before I was hooked. My passion is conservation and taxonomy, and much of my work has been directed towards developing a better understanding of the threats that affect Australia's invertebrates. I identify and describe the species most likely to be at risk of extinction, and work out what we can do to protect them. Without a doubt a favourite part of my job is field work, when I get to track down spiders in all sorts of weird and wonderful places.

On this particular trip my mission was to head to the Nullarbor Plain in South Australia, in search of a spider that had captured my imagination for years. I had been invited on an expedition led by the Australian Government's Bush Blitz initiative, which offered the opportunity to explore the limestone caves that dot the Nullarbor and document the invertebrates living in them, find new species and learn about threats to them. The Nullarbor is one of Australia's iconic landscapes, a vast and mostly treeless plain. It is embedded in the nation's folklore (just look up 'Nullarbor nymph' to see what I mean) and is an embodiment of the tough and uncompromising spirit of the Australian outback. This reputation places it firmly on the bucket list of many a traveller. The name Nullarbor translates to 'no trees', but despite its apparent emptiness, it is an area of incredible importance for biodiversity, endemism (species found nowhere else) and geology, most of which is hidden deep beneath the ground. The Nullarbor is home to the world's largest contiguous area of limestone caves. These caves were once an ancient drainage network, formed when the climate was wetter, and have been inactive since the early Pleistocene and the aridification of Australia.

Caves, with their low light levels, near-uniform temperature and humidity represent extreme natural systems. While many organisms that live underground are also able to survive on the surface, some species, known as troglobites, have become so specialised to living in caves that they are only able to survive underground. Troglobites often have adaptations to life in the dark. When no light penetrates the darkness, eyes are useless – troglobitic species often lack eyes and body pigment, and have elongated legs and appendages for sensing things in the dark cave world. One such group are the blind cave spiders of the genus *Troglodiplura*, an iconic and enigmatic group. Most species are only known from fragments of dead spiders found in caves and we know nothing about their ecology or biology. They are only known from the caves of the Nullarbor, and most species look like they may be endemic to a single cave. These animals were the focus of my mission.

The journey out to the caves – bouncing around in our four-wheel drive, along small and poorly marked tracks, trying to avoid the notorious blue bush (a puncture waiting to happen) – was a good introduction to the testing and uncompromising Nullarbor landscape. We finally made it to our camping spot for the night and set up camp next to one of the caves, pitching our tents on the red dirt, finding a space between the short, spiky, grey-blue stubby chenopod plants that dominate the area. Our camp was located next to tomorrow's cave, whose entrance comprised a surprisingly small and unassuming, but ominously dark, hole in the ground. As the sun slowly set, stretching out over the vast flat horizon, the transition of colour from golden to orange, yellow and pink was breathtaking. One of those 'if someone had painted it, you wouldn't believe the colours were real' kinds of sunsets.

With the darkness, we set out to do some invertebrate hunting. Night is one of my favourite times to look for invertebrates, as you get to see many species that are hidden away during the day and it is a perfect time to watch some interesting behaviours. A lot of large invertebrates were on the go that night. Spotting movement a few metres away, I moved closer and quietly watched as a very large and beautifully coloured red and orange centipede of the family Scolopendridae slowly walked along, poking its head into spider burrows until it found an occupied one and headed down for a feast. Shining my torch around to catch the glimmer of light reflected from wolf spiders' eyes, the plains looked like a field of stars, matching the glittering sky above. The wolf spiders were out in force, searching for prey and for mates. Hidden among the shrubs, I spotted a different sort of beauty, a large female spider from the genus *Miturga*, a member of the prowling spider family. She was sporting some impressive stripes on her body, typical of the genus, which gives it the common name of 'racing stripe' spiders. These large spiders are nomadic, ground-living hunters found across much of Australia. As with most of Australia's invertebrates, many are undescribed species. A nice find for the evening! Back to the camp and after some food (I firmly believe that the best-tasting food is camping food), I went to bed feeling both anticipation and apprehension about what tomorrow's descent might hold.

Some night-time thoughts: 'This is so exciting! I can't wait to see what we find down there. I really hope we find *Troglodiplura*, but what if I feel claustrophobic, panic and get stuck? What if I get scared of heights and get stuck? Can I do this? My camp mattress is uncomfortable! I can't wait to see what we find! But what if I get *stuck*?'

I woke up early in the morning. It was a bright and cold day, the sky a deep blue. Looking out over the vast landscape in the cool morning light and then looking down into the dark deep hole in the ground, my stomach gave a small lurch. I *think* I can do this. Before long we were strapped into our caving harnesses, fitted with a helmet (mine a cheery orange) and a headlamp and we were ready to go. As the cave guides lowered the climbing ladder down into the hole and secured it, we watched on, backpacks full of collecting vials, pencils and the all-important field notebook, and my mind full of thoughts and niggling doubts. One of the guides climbed down the 20 metres of ladder into the cave, while the other waited at the top to help us. Soon, it was my turn to descend.

The ladder was one of those narrow wire contraptions. About as wide as a large boot. As I climbed onto it, hovering above the dark abyss, my knees felt shaky and my hands were sweaty and feeble. 'Don't look down' I thought to myself. I descended slowly through the hole, the ladder wobbling precariously. I knew I was completely safe as the harness would catch me if anything went wrong, but try telling my knees that! After what felt like ages of slow careful descending (but was probably only a couple of minutes), my feet touched solid ground. 'And breathe'.

While waiting for the others to descend, I went to explore. We were in a cavernous chamber. The air was still, and very very old. The beam from my headlamp picked out details of the cave walls, sloping off in the distance down to the floor. As I looked back, the wire ladder hanging through the narrow hole in the cave roof seemed impossibly high and long; the person making their way down looked small and vulnerable, suspended high in the air. I turned around and scanned the cave with the beam of my headlamp, spotting a large spider moving along the ground. With a leap of excitement, I went to investigate. It was a racing stripe spider from the same genus as the one I found above ground the night before, but this female was much paler and lacked her relative's racing stripes. Alas, she definitely had eyes. Not quite the blind cave spiders of my dreams, but still, another good find.

With everyone safely on the floor of the cave we set off, exploring out to the 'dark zone' where no natural light reaches. If you were to turn off your light here, your eyes would never adjust to the darkness. It is here that we might find the animals that I longed to see, the species that have existed for millions of years away from the light. Existed here for so long that they have lost their eyes and the colour from their bodies, that their limbs have become elongated to be better able to sense things (including prey) in the pitch-black environment around them. What is the point of having eyes or colour if you can't see, and if no one can see you?

We moved carefully, trying to stand in the footprints made by the person in front, to minimise the marks we were leaving in the fine sediment. Just by being here we were altering this environment, marking it, changing it – as humans seem to change everything they touch.

The entrance to this cave – the hole in the roof – was effectively like a large trap. Evidence of the unfortunate animals that had fallen into the cave and could not escape was plain to see. I came across an intact dingo lying on its side, its fur looking soft and still sandy coloured. It could have just fallen asleep. Right next to it was a hole in the sediment, with clear scratch marks, where presumably the animal had desperately tried to find water or dig its way out. The dingo was dried and mummified but looked very similar to the way it had on the day that it lay down for the last time. This individual may have been lying there for 100 years, or perhaps over 1,000 years. The cave environment preserves bodies even after death. Towards the rear of this magnificent chamber the cave split into two much smaller passageways. Time for some crawling through tight spaces. The doubt reignited: 'Can I do this?'

The cave guide went first. I followed, scrabbling up over a wall of large boulders. The gap between the roof and the boulders became narrower and narrower, until I was shimmying on my stomach to fit through, pushing myself along with my toes, grazing my knees painfully on the rocks and with the solid heavy roof brushing against my helmet. Strangely, I wasn't scared. Sure, I felt a bit anxious, but also capable, and importantly I had complete trust in the cave guide. Eventually, the squeezing, bruises and perseverance paid off and the boulder floor started dropping away from the cave roof, leaving us with enough space to crawl, then to stoop and then to stand. We were through to a second large chamber. Around the edges the floor consisted of a fine orange sediment, with cracks in it where water must have collected at some point before drying up. The centre portion of the chamber consisted of a huge pile of large white boulders, towering to the ceiling. We set off to explore.

In the light of my headlamp, I spotted a dark shape on one of the rock surfaces. I was immediately on alert. Moving closer, I saw that it was a very large spider. My heart leapt in excitement. However, the spider was oddly still. It soon became obvious that it was dead – preserved by mummification just like the dingo had been. It was female and, judging by its size, probably mature. My heart did another somersault ... Where a spider's eyes usually occur, this spider had just a smooth section of carapace. I had found an eyeless spider! My celebration was short-lived because at that moment the cave guide shouted from the other side of the rock pile. 'Jess, you might want to see this'. As quickly as I could, which didn't feel very quick at all, I scrabbled up and over the rocks. There, on the underside of a large overhanging section of boulder, was another very large and very eyeless spider. This one was most definitely alive! My heart was thumping, she was beautiful and strange. She was unlike any surface spider. Her movements were slow and purposeful, almost robotic. Her limbs stretched out in front of her, feeling her way as she moved. Her body was large but graceful, a silvery brown colour, and her limbs were very elongate. As far as we could tell, we were the first people to have ever seen a live adult of this species; previously, adults were only known from fragments of dead spider. To see her in her cave environment, to watch this rare and unique animal move around, was mind-blowing! I was completely in awe. In that chamber we managed to find several juveniles and another two mature females; we also found several bits of dead males. We took many notes and images to record the spiders and their habitat. We were in our element.

On the journey back out we were all buzzing. The climb up the ladder, something that I had been dreading, felt easy, fun, exciting. We'd found the elusive high that field biologists love and strive for. Bolstered by our success, once on the surface we assessed our maps and, using the expert knowledge of the cave guides, picked some caves that shared the same features as the one we'd just been in. We jumped into our vehicles and headed to the next cave, once more bouncing along the rough winding tracks of the Nullarbor, full of anticipation.

The next cave had a walk-in entrance, much easier to access. From the maps and the cave geology and structure, it looked good for our spiders. We were hopeful, but as we slowly made our way deeper into the cave, out of the light zone, through the twilight zone and into complete darkness, we could smell a distinct difference between this cave and the previous one. Foxes. The smell became more and more intense. Through the beam of our headlamps, we noticed fox dens dug into the sediment and scats everywhere. In the dank cave air, the smell was almost overpowering. We surveyed the cave for an hour or two, searching thoroughly, but no spiders were found, even ones with eyes. We do not know whether blind cave spiders ever used this cave. If they had, we do not know whether the foxes affected them. However, I do know that foxes eat practically anything and that our large, slow-moving spiders would likely be very vulnerable to an introduced predator like a fox.

We visited several other caves on that trip, some pristine. Sadly, the more accessible ones often had evidence of human thoughtlessness and destruction. Discarded beer cans, stalactites and stalagmites snapped off, and dingo skeletons with their heads removed, presumably by people wanting trophies from their trip. We did not find any more blind cave spiders on that trip, but I have a strong feeling they are out there. Hopefully we can eventually map their distribution, so we can find ways to protect them and safeguard these large blind apex predators of the Nullarbor cave systems.

There is something special about being in a cave, which is connected to their antiquity and the divorce from the frenetic experiences of modern existence that being underground brings. It's about the way your line of vision is limited to a narrow beam from your headlamp, which makes lit features jump out and hides everything peripheral. It's about experiencing a world that is largely untouched by modern humans, which is delicate and fragile, ancient and proud. Caves are especially vulnerable to climate change and to threats posed by development, mining, incursion by feral animals and by us – the beer-can discarding, stalactite-snapping, dingo-head stealing animals. Yet we are all also custodians of these ancient places and the unique species that live within them.

On my long journey home from the Nullarbor, I contemplated the trip. I felt a different person from the one who had lain awake, worrying in her tent. I had been scared, I had doubted myself, I didn't think I could do it. By embracing the challenges, putting myself firmly out of my comfort zone, saying yes and taking risks, I learnt that not only was I strong enough but also that I loved doing it. Most field work is uncomfortable to some extent. You often have to rough it, take on challenges, get wet, mucky, hot, cold and be able to persevere despite your discomfort. If you can do it, the rewards for yourself and the species you are working to understand and to protect can be immeasurable.

Acknowledgements

I would like to acknowledge the Far West Coast people, including the Kokatha, Mirning, Wirangu, Yalata and Maralinga Tjarutja (Oak Valley) peoples, who are the Traditional Owners of the land in which we conducted our surveys. I also am extremely grateful to the Cultural Monitors who accompanied us on these trips. I would like to thank the Australian Government's Bush Blitz initiative, which funded this exciting research. Big thanks to cave guides Steve Milner and Andrew Stempel, for their expertise, professionalism and good company. Thanks also to Matt Shaw from the South Australian Museum for sharing his expertise and my enthusiasm for everything invertebrate.

11

Some cockatoos I've met

Erika M. Roper

I grew up in a house with an old disintegrating poster of the 'Parrots of Australia' stuck on the back of the bathroom door. Over the years the colours faded from steam and light, and it developed more and more tears. Eventually, it disappeared when my parents extended the house and built a new bathroom. To this day my dad insists that the many hours I spent looking at the poster are why I love birds in general and parrots specifically, and why I've ended up as a parrot researcher and cockatoo expert. No one in my very small extended family is a scientist or ecologist, but we still had a bird book, an old Simpson and Day field guide with a very ratty cover. I remember watching the crimson and eastern rosellas in the yard, then looking through the field guide with my mum to work out what they were and why they were different from each other. Coincidentally, my Honours project was on crimson rosellas and how they communicate with each other.

Of all the parrots in the book and on the poster, the group that captured my heart and mind with their zygodactyl feet was the black-cockatoos and their cousins, the gang-gang cockatoo and the palm cockatoo. Maybe it was because they were so different from all the rest of Australia's parrots. They are the largest of the Australian parrots and cockatoos, after all. Their colours are muted and monochrome, shades of grey and shiny black. A different colour palette from those of many of our other colourful parrots. They also have endearing personalities, although I didn't know that yet. There's something about black-cockatoos that captivates people around Australia and the world, no matter their background. Everyone has a black-cockatoo story. Here are some of mine.

My hometown is on Ngunnawal and Ngambri Country near Canberra, in the Southern Tablelands of New South Wales, south-eastern Australia, close to an area that has been used for pine plantations for many decades. Many pine trees are also planted on properties around town as windbreaks, including ours. It's a commonly known fact that yellow-tailed black-cockatoos eat pine seeds. I'm not sure if we know exactly when they started exploiting pine trees as a food source, but it probably began when their native food sources like banksia woodlands were cleared, and pine plantations were planted in their place. People say that black-cockatoos come when it's going to rain, but in my town that's not true. They come when the pine seed is ready to eat. As a kid I would hear them coming, calling as they flew into town over the hills with their whistling squeaky 'wee-loo', followed by chattering as they landed in the trees and settled in to feed. I'd run outside, hoping that they had chosen our trees to feed in that day so I could watch them – though my parents weren't too fond of them dropping pine cones on the car!

These days I work on threatened orchids and many of the sites I visit have resident yellow-tails. Many orchids are super tiny and very cryptic, so orchid work requires slow and careful walking, spending time looking at the ground to ensure that you don't accidentally step on any leaves or flowers. Slowly creeping around the bush while peering at orchids makes you appear very non-threatening to wildlife, including black-cockies. At one site, where I was looking for a threatened orchid in dwarf she-oak heath, the resident cockatoo family landed to feed in candlestick banksia bushes not 10 metres from me, completely unfazed by my presence. Another site that I regularly visit was severely burnt in the 2019–2020 bushfires. It is difficult to find words to describe the joy I felt at hearing yellow-tails calling there for the first time in the three years after the fires. Seeing a flock of 50 birds fly in over the still-charred tree trunks, their distinctive slow elegant wingbeats identifying them even if their call did not, brought a huge smile to my face and signalled that recovery was finally underway.

Yellow-tails still show up in my parents' yard, feeding in the pine trees and sometimes ripping into the trunks of wattles to find grubs ('grubbing'). Often I'll get a text message or a call from my dad saying, 'Your friends are here', and I'll know exactly what he means. Recently, my younger brother visited my parents and sent me a video of a very special (male) yellow-tail from the garden. Instead of being the usual solid dusty black colour with yellow tail panels, this cockatoo had a genetic mutation that caused many of his black feathers to be yellow, resulting in a mottled and speckled yellow and black bird. I looked for him the next time I visited, without success. Maybe one day I'll find him.

It came as no surprise to anyone who knew me, when I moved to Perth to start a PhD on the forest red-tailed black-cockatoo; also known as the karak, in the language of the local Noongar people. The largest of the black-cockatoos, the red-tailed black-cockatoo has five subspecies found across much of Australia, except the south-eastern section of the continent and Tasmania. The karak is one of these subspecies and is endemic to the tall forests of south-west Western Australia. It is a threatened species, with as few as 10,000 remaining.

About 25 years ago, if you went looking for karak in Perth, you probably wouldn't have found any as they were restricted to the jarrah forest, hence their English common name. Back then, they lived in the forest, feeding on the seeds of marri and jarrah gumnuts and nesting in the big ancient marri trees. They would drink from rainwater pools on granite outcrops, creeks and dams on forest properties. However, as free-standing water became scarce in the drought, some individuals may have left the forest to visit the edge of the city looking for water. There they discovered introduced cape lilac trees, which were full of delicious seed-filled berries that nothing else ate. Over the next 20 years these birds continued to visit the city and when I moved to Perth to start researching them, about 350 karak were in the city, feeding on cape lilac throughout the year. By the time I left Perth five years later about 3,500 karak visited the city, as word of a good food source spread exponentially through the population. I wanted to find out why. What was so special about cape lilac compared to the karak's traditional food of gumnuts? To do this I had to find cockatoos and spy on them, in the forest and in the city.

How exactly do you spy on a cockatoo? The most efficient way to locate cockatoos is to drive around slowly with the windows down, listening for either their distinctive calls or the sound of crunching as they chew through different fruits. After five years of cockatoo spying I could often identify what tree species they were feeding in before I saw them, just by the sound of the crunching. However, aside from their crunching, karak are very quiet when they feed, so unless they have a juvenile (juvie) with them you're unlikely to hear them call. If they do have a juvie, all bets are off as the kid, like all youngsters, will be making a huge racket, constantly begging and generally making a nuisance of itself. Juvie karak have two main begging calls that I noticed in my time spying on cockatoo families: the very insistent and urgent 'FEED ME NOW' call, and a softer, more plaintive 'please give me attention'. These two calls can also be distinguished by slight changes in the juvie's posture, with hangry juvies likely to be more hunched with their head tilted up, or possibly also awkwardly chasing their parent around the tree for food. Parent cockatoos must have a super-cockatoo ability to tune it out, as they just go about their business, processing fruits and eating seeds, ignoring the incessant demands of their child until they are ready to feed them (known as 'allofeeding').

Younger juvies will just sit and beg, but as they get older they watch their parents to learn how to process fruits and start to practise themselves. The learning process can lead to some entertaining observations. Just like human babies, juvie cockatoos are very clumsy, both at handling food items and at moving – in their case, navigating through the branches of trees. I've often seen them follow their parents through trees, climbing like all parrots do using both feet and their beak, but because they are still learning how to control two legs, two wings and a beak they are very uncoordinated, and often lose their balance and wobble around. If they manage to pick a fruit, chances are they will drop it. Once they get more skilled at keeping hold of the fruit it takes them a long time to figure out how to process it and extract the seed. Even then they often accidentally drop the fruit partway through. It must be frustrating to be a juvenile cockatoo! Fortunately juvie karak are dependent on their parents for up to two years, so they have a lot of time to figure out how to be a cockatoo.

Many times I've seen juvies try to steal food from a parent. One of the most interesting instances of this happening was late one afternoon in the jarrah forest to the east of Perth. I was watching a family group of karak feeding in a marri, happily chewing away on the large gumnuts. Suddenly another family group flew in, calling to announce their arrival.

A juvenile from the second group made a beeline to the female in the first group and attempted to steal the gumnut she was eating. A brief tug-of-war ensued. The juvenile won. It walked a few steps away and proceeded to drop the gumnut without eating it. To this day I'm not sure what exactly happened.

Cockatoos are smart. I swear they knew when I was watching them and would turn their backs on me so that I couldn't properly record them feeding. They also had uncanny aim when dropping their finished gumnuts; a surprisingly large number of them seemed to land right on my head, with force! I also learnt that cockatoos really like big trees, especially in urban areas. I spent a lot of time driving around the older suburbs of Perth listening for cockatoos, before setting up to record them in street trees outside people's houses and sometimes in people's backyards. It was always a little awkward when the occupants or their neighbours arrived home to find me standing outside with a video camera seemingly pointed at their house.

Urban areas with lots of really large trees, aside from bushland reserves, include parks, playgrounds, sports fields and schools. All places that attract cockatoos. And children. Once I realised the correlation between preferred cockatoo habitat and preferred children habitat I timed my field trips to these locations to occur when children would not be present. Apparently a person with surveillance equipment around children makes people twitchy! Fortunately no one called the police, though I did speak to a few principals and some parents. Once I explained what I was doing, pointed out the cockatoos and did some impromptu science communication most people were happy.

On one of my first urban cockatoo spying missions, I was driving around the more central suburbs of Perth, windows down, ear cocked, looking and listening for cockies. I had been searching for quite a while as I wasn't yet familiar with the cockatoos' behaviour or with Perth, so I hadn't figured out their daily routines or the lay of the land yet. Having moved to Perth from 'over east' (as they say in Western Australia) I was also not aware that there are some locations in Perth where you just don't go.

Finally, I saw a pair of cockatoos in the distance! I tracked them through the suburbs and eventually they landed in a big tree at the back of a hospital, which coincidentally was just down the road from their night roost. This hospital looked very small and unassuming, with no gates or signs saying 'KEEP OUT' or 'NO ENTRY'. Just a sign saying 'HOSPITAL', so in I drove. I followed the roads until I got to a parking area close to the tree in which the cockatoos had started feeding. I parked, hauled out my equipment, set up and started recording. It was early in my research, so I was testing out methods and equipment. I had a video camera on a tripod, a camera with telephoto lens, binoculars, a clipboard with datasheets and a notebook. As a poor PhD student, I was also in my own unmarked car and not wearing any sort of uniform, so I just looked like a scruffy random person with surveillance equipment. Personally, I thought I looked like a totally normal person doing completely normal things. However, the two burly security guards who screamed up in their car, lights flashing, did not. It turned out that this small unassuming hospital with no signage was actually a high-security mental health facility. I had to do some very fast talking to explain what I was doing there. I think the fact that I had an eastern states accent (yes, it's a thing) helped convince them that I was an innocent and ignorant Perth newbie. Fortunately, the cockatoos were feeding in a tree away from any buildings because I'm sure that if my camera was pointed even remotely at a building I would have been arrested. Instead, I was escorted from the premises after all my details, including my car registration, were recorded.

When I told my Perth friends about my adventure, that evening at the pub, they all cracked up laughing and said 'Of course you shouldn't go in there!' Needless to say, when I used a random number generator to choose some spatial points for habitat surveys a few months later and one of them was smack bang in the middle of the hospital's grounds, I strategically moved the location, just a bit.

In the forest you still have to worry about people. I often had to leave an area sooner than I'd planned because I heard gunshots coming progressively closer to where I was recording. On one occasion I couldn't leave, and my super sneaking skills came in very handy. I was recording cockatoos deep in the forest. Out of nowhere a man drove up a narrow track and started illegally harvesting firewood about 30 metres away from me. He scared the cockatoos away, but I had to stay still and hide behind a tree for several hours until he drove away again, illegal harvest loaded on the back of his ute. I was alone, in the middle of nowhere, and I did not want to confront a man with a chainsaw doing something illegal.

That evening at the local pub, while I was eating the only vegetarian dish on the menu (surprisingly tasty deconstructed nachos that I ate too many times over the years), I mentioned the incident to the locals. Unfortunately, they confirmed that illegal firewood harvesting was not an uncommon occurrence in the area. I spent many nights in that pub, entering my data, backing up my videos and chatting to the locals about cockatoos and other birds, before heading back to the bush to sleep in my tent. One old codger liked to tell me how when he was young they'd routinely catch 28s (a colloquial term for Australian ringneck parrots) for dinner. According to him, the reason they were called 28s was because you needed 28 parrots to make a pie to feed a family. You don't see 28s around much anymore in Western Australia, especially not in urban areas where the introduced rainbow lorikeets have taken over, so it's a good thing that parrot pie is off the menu.

One constant in my life is that cockatoos will distract me. Recently, I was in the Northern Territory, or the Top End as the northern part of the Territory is known, driving down a back road from the tiny town of Pine Creek and looking for hooded parrots. I was heading to a sewage treatment plant and adjacent cemetery that had reliable recent records, because (a) why wouldn't you put your poo farm and bone orchard next to each other? And (b) birds love both of those places. Lo and behold, when I arrived a large group of cockatoos was foraging 'normally' in the trees and, more unusually, on the ground. What's a cockatoo researcher to do but spend several hours creeping around spying on the cockatoos, trying to work out what they were doing on the ground? Unfortunately, they kept flying away before I could get a good look at what they were doing, but I assume they were eating something. I had a good wander around the area after they left but I couldn't find any evidence of what they were feeding on. To this day it's a puzzle!

As I neared the end of my lap of the poo farm I came across a large pile of fresh buffalo scat, which made my heart stop for a moment. At that point I realised that getting distracted by cockatoos in the Top End was risky and maybe not the best life choice. I carefully and quickly made my way back to the car, avoiding the temptations of the rest of the cockatoos, and jumped back into the relative safety of the hire car. Whether a Toyota Corolla could hold up to a buffalo is not something I wanted to find out, so I didn't hang around.

As I left the poo farm I encountered an old hunter and his four dogs on a quad bike (yes, they were all on the bike). We had a good chat about birds, and he gave me tips about staying safe in the Territory, especially from buffalo and crocodiles. I learnt that if you threw coins at a buffalo they could hit the coins every time with the tip of their horns, that they are super sneaky and invisible in the bush, and that if you see one your best bet is to climb a tree and stay there forever. As for crocodiles, no matter how far away you are from a river, you're not far enough, and they will climb into the back of a ute to take your dog (RIP Rover).

Cockatoos find me wherever I am, even if I am not looking for them. Recently, I was driving to Griffith in the Riverina region of New South Wales to visit a good friend, stopped at a tiny town for a wee and what did I hear when I get out of the car but a red-tailed black-cockatoo calling? Bursting bladder forgotten, I grabbed my binoculars and camera because I was outside the known range of any of the red-tail subspecies, so I had to find this bird! After a fruitless 10 minutes searching all the trees in the immediate area, I realised that the calls were coming from the back of the pub. I popped my head into the pub and asked if they had a red-tailed black-cockatoo. The bartender seemed confused, so I explained that I recognised the call and that I researched them back in Perth. Anyway, the answer was yes, they kept one out the back, which made more sense than there being red-tails hundreds of kilometres out of their range.

These are just some of the many many cockatoos I've met in my life, and doubtless I will meet many more. I've also been fortunate enough to meet members of the other black-cockatoo species, Carnaby's (short-billed white-tailed black-cockatoo, Ngoolark), Baudin's (long-billed white-tailed black-cockatoo, Ngolak) and, last but not least, glossy black-cockatoos during a recent survey for eastern bristle-birds in Barren Grounds National Park. I had looked for glossies many times in the past and they always eluded me, so it was amazing to finally see them and hear their creaky call as they flew over the heathlands.

I can't finish this chapter without mentioning the cousins, the gang-gang and palm cockatoos. Often mistakenly called black-cockatoos, they are in fact cockatoos that happen to be black (or grey in the case of the gang-gang). The first time I saw gang-gangs was a nesting pair in Canberra, the female snoozing in a hollow, her boyfriend with his jaunty red cap and curly crest keeping watch above, on alert to make sure I wasn't up to anything nefarious. Their distinctive creaky-gate call is always a delight to hear. I am yet to see palm cockatoos, as they are restricted to the tippy-top part of Queensland, up in Cape York. Perhaps this is a good thing, as my cockatoo story is not over and there are so many more cockatoos for me to meet and many more adventures to be had.

Parrots and cockatoos are amazing adaptors and can be found all over Australia, including in towns and cities. I wouldn't be surprised if you have cockatoos in your backyard right now! They might be pruning some of your trees or providing free lawn aeration services. Everyone has a black-cockatoo story. What's yours?

Acknowledgements

Thank you to my parents, granny and family for supporting and indulging my parrot quests over the years. Thank you to all my friends in Perth and on the internet for their support during the PhD times and beyond. Thank you also to the hospital security guards for not arresting me. And a big thank you to the National Parks & Wildlife Service for producing the parrot poster (you should really consider doing a reprint).

What a botanist can learn from a dog

Laura M. Skates

This story is about a botanist exploring a remote part of northern Australia, in search of carnivorous plants. It's a story of adventure, spectacular landscapes, steep learning curves and extreme heat. It's also the story of a loyal dog named Bonnie. As her name suggests, Bonnie was a beautiful dog, with a tan-coloured coat, long legs, pointed alert ears, curious brown eyes and a heart of gold. However, before we get to Bonnie, and the lessons I learnt from her, let me tell you about my journey to the remote carnivorous plant hotspot known as the Kimberley.

***

It was 2015, and the first year of a long journey towards gaining my PhD. My thesis topic was a dream come true for me: the nutrition and ecology of carnivorous plants. I had always been fascinated by the natural world: from spectacularly coloured rocks to Australia's unique wildlife, to cloud formations in the sky – all of it fascinated me. Anyone who has spent time in the south-west corner of Western Australia, where I grew up, will know how easy it is to fall in love with the mighty karri forests, white sandy beaches and wetlands, granite outcrops, and bushland full of endemic (found nowhere else in the world) plants. I knew this place was special. However, I didn't realise that I was also living in one of the world's most diverse carnivorous plant hotspots until I was much older.

It was during my undergraduate degree, where I majored in botany and conservation biology, that my fascination with plants really grew. I became captivated by the incredible diversity of Australia's native flora, its unusual shapes and colours, and how it had adapted to survive and thrive in some of the trickiest, and often harshest, of circumstances.

When a plant needs water or food, or when it is ready to reproduce, it cannot simply uproot itself and move to find the resources it needs. For the most part, plants are quite literally rooted to their spot and they have to make do. I learnt how our native plants have adapted to all sorts of challenges – fire, salinity, drought and incredibly nutrient-poor soils. I learnt about plants that have adapted to those challenges and turned them into opportunities – the plants that use smoke and heat from fire to release and germinate their seeds (like hakea), the plants that store water in their tissues to save for a not-so-rainy day (like the boab tree), and plants that find nutrients in the strangest of places. Some hungry plants are parasites that feed on other plants growing around them (such as the spectacular Western Australian Christmas tree). Others use their roots to form below-ground friendships with fungi for sustenance (as is the case for many orchid species). And then there are the carnivorous plants, which use their leaves to capture unsuspecting prey and make a meal of it.

I was fascinated by all of them, but it was the final group of plants that really intrigued me. The carnivorous plants of the world have flipped the tables on the supposed natural order of life, becoming not only photosynthesisers and primary producers, but carnivores too. I still recall the first time I saw an elusive carnivorous plant, the Albany pitcher plant (*Cephalotus follicularis*), in the wild. It was during a field trip to Albany on the south coast of Western Australia, where we were listening for native frogs in a swamp. Standing out from the dark peaty soil were bright pops of green. Closer inspection revealed that they were not frogs, but leaves forming tiny cup-like traps to capture even tinier insect prey. I remember inspecting these weird leaf formations and thinking 'How is this plant real?' I would soon learn that this species was endemic and unique, being the only species in its genus, and the only genus in its family. Yet this one-of-a-kind species was also somehow not unique, being remarkably similar in form to a group of tropical pitcher plants known as *Nepenthes* and the trumpet pitchers known as *Sarracenia*, which each evolved completely independently from the pitcher plant I saw before me (a phenomenon known as convergent evolution). Most fascinating of all, I also learnt that they were, of course, carnivorous. I had to learn more. At the end of my undergraduate degree I was lucky to land an Honours project studying native carnivorous plants. This work led to my PhD and my first field trip to the Kimberley.

*****

Travelling to the Kimberley was an adventure in itself. I was one member of a small field team, consisting of myself and two other researchers. We would be spending the week at a remote field station, where we planned to study different aspects of the local bushland. Much like the south-west corner of Australia that I'd grown up in, the Kimberley region is a hotspot for carnivorous plant diversity. It was a perfect place for me to study the nutrition and ecology of a wide array of carnivorous plants living alongside one another in their natural habitats.

The Kimberley is located around 2,000 kilometres north of my hometown, Perth. To make our way there, we took a flight from Perth to Broome then another from Broome to Kununurra. I remember stepping off the first plane in Broome and walking into a wall of intense heat. In Kununurra, I saw invasive cane toads for the first time, signs warning of crocodile-filled waters, and beautiful plants that I'd only ever seen at botanic gardens. We stayed there for one night, before heading off early the next morning.

The quickest way for us to get from Kununurra to the remote field station where we'd be spending the week was by mail plane. My preparation for the trip included packing a small bag of field clothes (hat, sunglasses and long-sleeved shirts for sun protection, and gaiters to protect my ankles from snake bite), filling a tub with equipment (including sample bags, silica gel and a large water bottle), and clearing enough space on my phone for all the photographs I knew I would take. Our bags of clothing and field equipment were weighed carefully before being loaded evenly onto a tiny Cessna Caravan airplane. I and the two other researchers were also weighed and loaded onto the plane, alongside all the groceries and mail that were delivered to each of the remote field stations every Tuesday.

When I think back to my time in the Kimberley, one of the first things I remember is the view from that little plane. I imagined myself as a bird, flying high over the land, seeing the twists and turns of rivers and the mosaic of greens, browns and blues. That memory also reminded me of my grandfather who was a surveyor. He would have flown in planes just like the tiny Cessna over landscapes just like the one I was flying over. He had passed away the year before, so it felt to me like he would have been watching in that moment, enjoying the beautiful views with me.

Along the way we made a few stops at remote stations, often meeting people near the runway ready to collect that week's delivery. When we eventually arrived at our destination, we hopped off the plane, collected all our belongings and met the station caretakers. It was then that I met Bonnie the dog.

***

It bears repeating that Bonnie was a gorgeous dog. She greeted us with enthusiasm (much tail wagging and kisses hello) and I quickly learnt that she was always ready for an adventure. We were going to be staying at the field station for one week exactly. Provided the weather conditions were suitable for landing, the mail plane would be back the following Tuesday to drop off deliveries and pick us up. Until then, we had work to do.

Every day, we would wake up early in our dongas (small portable rooms), get ready and head out into the bushland surrounding the station homestead. Using the station's set of bright red quad bikes, we followed tracks that others had made through the bush. Although it was early, temperatures were already climbing and the cool breeze as we rode was a relief. Right there next to us, ready to go, was Bonnie.

We set out in a conga line of four: three quad bikes and one dog. As we drove through the bush, Bonnie would run to the front of the line and lead the way. After all, she had lived at this station for quite a while and knew the area like the back of her paw. So, of course she'd know the best spots to go and the best routes to get there.

For the person riding last in the queue of quad bikes, it was important to maintain a safe distance while also keeping up with the quad bikes in front. Every so often, Bonnie would run from the front of the line to the back of her 'pack', checking in. If you were last in line Bonnie would run for a while alongside you, keeping pace easily while also keeping you company. Once she was satisfied that you were well, she would run off to the front to lead the way again. Bonnie would do these checks multiple times on our journey, running back and forth, simultaneously guiding us through the bush and ensuring that every member of the team was accounted for.

Eventually, we would find our way to our field sites. We'd hop off the quad bikes, ready our sampling equipment and begin to search for our targets. It may come as a surprise, but carnivorous plants are often beautiful and delicate. Not at all the monstrous people-eating creatures you might expect from their portrayals in popular culture! From Audrey II in the 'Little Shop of Horrors' to the piranha plants of 'Super Mario' world, carnivorous plants are often used as a horror trope and yet I can never think of them that way.

The carnivorous plants I was after can sometimes be tricky to spot, as their delicate stems and leaves are well camouflaged among the foliage of other plants. The rainbow plants (*Byblis*) have long slender leaves covered in sticky hairs, with large five-petalled purple flowers sporting bright yellow centres. As their name suggests, the sticky leaves of these plants have a shimmering rainbow-like quality when seen in the sunshine. The sundews (*Drosera*) are similarly covered with sticky dewy leaves, and their shape and size can range from tiny rosettes of red–green leaves lying flat on the ground to tall herbs standing vertical with sticky tentacle-covered leaves outstretched. The bladderworts (*Utricularia*) don't have a pretty name but do have quite lovely flowers on long stalks, with specialised suctioning leaves that usually sit either below-ground (terrestrial plants) or float in water (aquatic members of this genus).

On countless occasions, when I'd been searching for carnivorous plants through the bush for what felt like ages, I'd finally spot one. Triumphant with discovery, I'd look up – only to realise that the sparkling plants were all around me. My eyes just hadn't picked them out from the backdrop of the Australian bush before, but once I'd seen one set of sticky dewdrop-covered leaves, I saw them everywhere. Occasionally I would be lucky and the bright pink of a *Drosera fragrans* flower or the purple flower of a *Byblis* would stand out against the green surrounds. Often, I'd need to kneel on the ground, pushing vegetation aside to find some of the smaller sundew species peeking through, tiny insects caught on their sticky leaves.

When I found what I was looking for, I would set up a makeshift workspace with my box of sampling equipment and notebook at the ready. Under scientific collection licences, I would carefully take small samples of leaves from the carnivorous plants and surrounding non-carnivorous plants and collect insect prey and soil. All of these samples would be dried and prepared for stable isotope analysis, to understand how much of the plant's nutrition had come from the unlucky prey captured within its leaves.

Every so often, we would encounter wildlife among the bushland, including unfamiliar bird calls and insects. I still recall the excitement and awe of stumbling upon a pile of white shells and bone fragments laid out at the entrance to a U-shaped structure of sticks, expertly created by an unseen great bowerbird. Another time, a giant monitor lizard (or goanna) appeared before us. All I could think was that I'd never seen a lizard that big before. Bonnie kept a watchful eye on the monitor as we continued along the track, protective of her pack. We'd take care of her too, making sure that shade and a creek were nearby, or sharing water from our bottles. While we worked, Bonnie would be nearby sniffing around and exploring, periodically checking that we were all still okay. She'd eventually find a shady spot under a tree and wait patiently.

Once we'd finished our field work for the day, we'd prepare for the ride back to the station. We would pack up our equipment and samples into tubs, strap them to our quad bikes and jump on. By this time, we were all pretty exhausted, but none more so than Bonnie. After all, she had run laps around us all day! Knowing it was the final part of the journey, she would choose one person and look up at them, very politely and silently requesting a lift home with her big brown eyes. If you were the chosen chauffeur, she'd jump up and settle into the tray behind you, safe and secure. With such precious cargo, we'd drive very slowly and carefully back to the station as the sun started to lower in the sky.

***

The lessons we learn in life can come from surprising places and people. Sometimes, they can even come from a dog. A PhD itself is a journey of learning. Not only learning about your field of research, diving into the topic and contributing new knowledge through the process, but also learning about how to be a researcher, how to lead a project, and how to sustain your energy for that years-long marathon.

During that week in the Kimberley, I expected to learn a lot about carnivorous plants. What I didn't expect was how much I would learn from our canine companion. Through her actions, Bonnie taught me some important lessons about leadership, teamwork and caring for others.

Bonnie was a good leader. She had the confidence to take charge and lead the way, taking it upon herself to guide us through the bush trails in the way that she knew best. She demonstrated that being a good leader isn't just about being at the front of the pack. It's also about empowering and caring for the team around you and checking in with every team mate to ensure that they're all okay. At the end of the day, she taught me that good leaders also take care of themselves. They know when to ask for help, and when to rest. Most of all, she reminded me of the importance of greeting each day with enthusiasm and an adventurous spirit.

***

Two years later, I had the joy of meeting Bonnie again. I was a couple years into my PhD, and excited to spend another week in the Kimberley searching for carnivorous plants. Even though Bonnie and I had only met for one week, two years prior, I like to think that she remembered me on my return, even just a little bit. I've thought about Bonnie a lot since those trips and held onto all those lessons she taught me. As it turns out, a botanist can learn quite a bit from a dog.

Acknowledgements

I acknowledge the Traditional Custodians of the lands upon which this work took place, the Whadjuk Noongar people (Perth) and the Ngarinjin, Miwa and Gamberre people (Kimberley). My field work experiences in the Kimberley were made possible thanks to funding from the Friends of Kings Park and Kimberley Society. Thank you to the Theda station caretakers for their kind hospitality, and to my field work companions Dr Adam Cross, Dr Christina Birnbaum and Dr Todd Buters, and my PhD supervisors Prof. Gerhard Gebauer, Prof. Kingsley Dixon, Dr Jason Stevens, Dr Adam Cross and Prof. Erik Veneklaas. Finally, thank you to Helen Waudby, Sara Halas, Joanne Castelli, and Jeff Skates for their kind support and helpful edits of this chapter. And, of course, thank you to Bonnie.

13
Bears, drugs and guns

Andrew E. Derocher

From an early age, I had settled on a career working outside, though studying wildlife as a career wasn't something I had ever imagined. And while there was no such history, nor strong links to university education, in my family, my passion for the outdoors slowly morphed into a love of wildlife. Despite growing up in a temperate rainforest on the west coast of Canada and training as a forest biologist, I've spent most of my career studying polar bears in the cold treeless seascape of the Arctic.

I'm an accidental academic and have found that a career as a field researcher in wildlife biology often takes a serendipitous path. However, I am in my element in the outdoors and field work remains a driving force in my career. In the 1980s, I was studying grizzly bears in coastal British Columbia on the Kimsquit River. As I donned my heavy-duty rain gear to visit feeding sites, my boss asked what I planned to do for the winter, as my current position focused on tracking the bears until they hibernated. When I said that I planned to look for a graduate placement, he mentioned that he knew someone who studied polar bears and could help me connect with them if I was interested.

I hadn't planned on working abroad, let alone in the Arctic (I abhor being cold), but polar bears intrigued me. The contact was fortuitous, and I went on to complete two graduate degrees on the species. Today, my research focuses on polar bear population ecology and how they are responding to climate change. Thanks to much luck, some hard work and a focus on good science, I've completed many field projects over the decades, spanning both North and South Poles. Each field trip is a bit different – some are memorable for good reasons and others for less positive ones, but they have all taught me something.

Most importantly, research should never be an 'adventure'. Adventure comes only to the poorly prepared and incautious. While good and bad luck happen, field work is part of your job – it's not recreation. It may not sound like much fun, but as the saying goes, there are 'old biologists and bold biologists, but precious few old, bold biologists'. Planning, equipment, emergency preparedness, contingency plans, constant monitoring of the weather, considerable patience and a healthy fear of death are critical components for an Arctic field program, and no doubt for other field work.

Planning is key to a successful field project and it should start early as numerous components must fall into place. For my research, animal care protocols need to be approved, permits obtained, helicopters booked, fuel cached, immobilising drugs ordered and more. Many of these elements take months, in some cases years, to set up. It took me over three years to get permission to sample polar bears in a high-level nature reserve in Svalbard, Norway. Never underestimate the power of bureaucracy to thwart your plans!

Unexpected logistical problems can be equally disruptive. I once lost two days of field work looking for a wayward fuel cache because the GPS coordinates hadn't been noted accurately during the previous summer. This situation wouldn't have been such an issue in mid-summer when the vegetation is no taller than your ankle, but in mid-winter, after months of blowing snow, the drums were lost under a blanket of white. Finding the fuel was critical as it was the link to our destination further north – the destination for which we'd waited three years for a permit. We spent hours walking a grid, spiking downwards with metal rods and hoping to bang a buried drum. Fortunately, a passing polar bear had smelled the drums and dug down to them, and we happened to arrive before the hole blew over with snow. We went on to sample the remote reserve and it was a highlight of my career – the island was like a polar bear condominium complex in spots, with numerous maternity dens snuggled close together. Needless to say, I had a discussion with my logistics people after that debacle! Careful marking of the GPS position and a simple signpost could have avoided the issue.

All in all, a happy, well fed and well rested crew will vastly outperform a disgruntled one. I've worked in field camps for over 45 years. A good field camp is a lot of fun with memorable experiences shared by colleagues and collaborators. On the flip side, a malfunctioning field camp causes stress and conflict. A field camp is a team and having everyone pulling the same way is critical. Stress is a common element of field research and adding to it doesn't make anyone a welcome camp member. My advice is simple: do more than your fair share of camp duties. Be the first one to dive in to cook, make coffee, wash dishes or clean up. Just one lazy or pessimistic crew person can cause a loss in productivity. If you're not already a great cook, learn a few dishes for carnivores, vegetarians and vegans and you'll do well. Ask about food restrictions, preferences or allergies. Field work is fuelled by good food. I've been in camps where vegans are limping along because of few good meal options and carnivores are longing for a pork chop. I'm no gourmet cook, but I can turn out meals that feed the masses no matter their preferences or limitations. Taking the time to ensure folks are well fed is a critical component of research.

Being confident with your gear is fundamental to safe and effective field research, particularly when working with potentially dangerous animals. For example, learning how to be reflexive with bear spray or a firearm is critically important in my line of research. More than one attempt to ward off problem bears has gone awry when the device shot at the bear to scare it away overshot and scared it forwards instead. Years ago, I was charged by a grizzly bear and had to fire a warning shot to scare it off. However, if the bear hadn't been deterred, and had attacked instead, I would not have had an opportunity for a second shot. In hindsight, I should have saved my one shot for protection and not used it as a deterrent, which is what I tell my field crew. While my research crew has extensive firearms training, little can prepare anyone for facing down a charging bear.

Study animals can be dangerous, and polar bears certainly are. I've never seen evidence that they actively hunt people, but they are curious about humans. I'm convinced that it's not the bear I'm trying to catch or have caught that is the dangerous one, it's the one I haven't seen yet that's potentially a problem. Polar bears have a contagious distribution: where there's one bear, there are likely two and where there are two bears ... You'd think that the vantage point of a helicopter would let you see every bear in an area, but polar bears have many hundred thousand years of evolution on their side. A whitish bear on a white background isn't obvious at the best of times, and often the bears fall asleep and get blown over by snow.

Like bears, pilots can also be overly curious. On one outing, we spotted the remains of a seal kill near a bear that we were in the process of catching. After the bear was drugged, we started taking measurements but, unbeknown to me, the pilot decided to go for a walk to find the seal kill. We collect specimens from the kills, and he figured it was a time-saving move. It was only as he was returning with a dead seal pup in hand that I saw the bear following behind him. Losing our pilot far from camp wasn't part of the plan, and thankfully some loud shouting and gesticulations convinced him to drop the seal and make for the helicopter. Fortunately, once the bear reached the seal it gave up the chase. If there's a moral to this story, it's that communication and knowing what's going on are vital. Had I known the pilot's intentions, I never would have let him wander off alone and unarmed.

Weather is the most likely reason for trouble during my field research. Often, weather forecasts are unpredictable, so a plan to overnight far from camp is routine and critical. While it might not seem overly sophisticated, we often carry garbage bags filled with rocks to throw on the ground as reference in case we hit a white-out, which is a far better option than jettisoning your gear. Field gear for the Arctic includes a lot of safety equipment, but cold-weather sleeping bags, sleeping mats, tent, stove and fuel to melt snow are essential. I've spent the night on the sea ice waiting for weather to clear. Having to set down on the sea ice for extended periods is not preferred but can happen. It's not fun, and curious polar bears are often about. The rule of three is useful to remember when planning for these types of situations: people can survive three minutes in cold water, three hours of exposure, three days without water and three weeks without food. It's obviously a rough index but it keeps things in context – you're unlikely to starve, at least, before help shows up.

Polar bear research typically happens on sunny calm days that aren't too cold. We need the sun to find their tracks and low winds so they aren't erased. It can't be too cold, or we risk the helicopter battery not having enough juice to start. My watch has a barometer – a change in pressure is a good indicator of a shift in conditions. An Arctic high usually means clear and cold. Falling pressure could be a storm. Conditions might be fine right in the middle of a low-pressure system, but often deteriorate quickly. While standard public forecasts can be helpful, I've found that the marine and aviation forecasts are more useful when planning a trip. In remote areas, weather forecasts aren't very accurate – not many people live in these areas and few resources are dedicated to making predictions. I once sat in a remote camp for three weeks waiting for fog to burn off so we could fly. Every day the weather office said it would clear by noon (maybe), but it was a long wait for that forecast to be correct. I learnt to read the weather by working with locals who knew what might be coming, but also, critically, by learning how to read weather charts myself. Learning how to read weather maps can save research funding and enable you to collect more data in a safer manner.

While helicopter maintenance is held to incredibly high standards, it's not infallible. I remember one instance when I caught a big adult male polar bear from the air and we had to land the nose of the helicopter just a few metres behind the bear because space was so tight. After processing the bear, we jumped back in the helicopter and the pilot hit the starter. Nothing happened. A few more tries and still nothing. Luckily, the helicopter technician was with us that day, as I'd offered him a spare seat so he could get out of camp. He figured out that the starter was broken and amazingly had brought a spare. Of course, some time had elapsed by the time we'd figured this out, and a huge groggy bear was beginning to wake up and give us some odd looks. We didn't have the right tools to install the replacement starter, so the three of us used our multitools to pull the broken bit from the existing starter. The technician installed the new one just as the bear was starting to stand. The drugs we use produce rather relaxed bears as they wake, but I was still pleased when the starter kicked in and the blades were spinning. Sometimes good luck counters bad!

The ability to be adaptable and patient during field work is a useful trait. During cold conditions, helicopters need to be plugged in to keep the engine warm and various oils sufficiently fluid to fire up. Having wolves play with the plug-in cord, and unplugging it in the process, once caused a delay of a few hours that I just had to wait out. Patience can also be applied proactively. On a polar bear research project in the Canadian High Arctic region, I held a permit to study both polar and grizzly bears.

Grizzlies are expanding in range, and it was an area where wild hybrids (animals that result from interbreeding between polar bears and grizzlies) had been recorded. We had just refuelled the helicopter and were searching the nearshore ice back to camp for bears when a large dark object appeared on the sea ice – it was a large adult male grizzly hunting in polar bear habitat.

I knew that bear would be a valuable data point, but a dark and foreboding sky between us and our campsite indicated a potential storm. Safety trumps all, so I made the call to head back to camp. We got about halfway there before we experienced white-out conditions: an incredibly dangerous situation where snow results in loss of visual horizon and pilots can instantly become disoriented and crash. Before things got that bad, we dropped to near sea level and were able to move from one safe landing site to another. Until we were eventually stuck, and couldn't move safely any further. We shut down the helicopter and called air traffic control by satellite phone, reporting that we were safe but waiting for the weather to change. A few hours later, conditions improved and we safely made our way to camp. Twelve months later, I was flying in the same area as the last year of the study when I saw what looked like the same grizzly very close to the original site where I had spotted it. This time, things worked out and the capture was a success. This big male hadn't fathered any hybrids with polar bears, based on the genetic analyses that followed, but it was still a valuable insight into an expanding population of grizzlies.

A difficult lesson to learn is when to call it a day or wrap up a season. Both are judgement calls and I've learnt the hard way that the 'last bear of the day' is sometimes best left as the 'first bear tomorrow'. When people are tired, cold and more interested in dinner than in collecting more data, it's a good time to stop for the day. Wildlife research is often dangerous, and tired people make poor decisions. Knowing your own limits and being able to assess, and respond to, the state of others in your field party is part of being safe. While pushing to meet a study's objectives is an admirable goal, doing so unsafely can quickly go sideways. Wildlife field researchers are usually in their chosen occupation because of their love of nature, or maybe the joy of discovery. Pushing the limits in the field is a good way to truncate your career.

Sometimes, despite good pre-planning and tested protocols, work doesn't go according to plan. Being as ready as possible for these situations is critical. We were studying female polar bears with newborn cubs emerging from their spring dens in the Norwegian Arctic in the late 1990s. I had located a mother and cub high up in a snow-covered valley. As per protocol, I darted the mother first as she moved down the valley, before approaching the cub. Spring cubs are small enough to catch by hand, but this one had a different plan and headed straight up a mountain, where finding it among the rocks would be tough. I decided to follow it and told the helicopter pilot to nose up to the cliff, so that I could climb out along the skid and catch the cub. As the helicopter dropped me off, I took one step and realised that the cliff was perfectly glazed by a recent (and aberrant) rainfall, and it was incredibly slippery. One false step and I would career down the hill. At that point, the mother, groggy but not asleep, spotted me between her and the cub, did what every good bear mother does and decided to protect her cub. I had my firearm, but it was not a comfortable situation to be in (sliding down a mountain, shooting a bear, or both). The pilot and crew initially didn't notice the mother's change of direction, but quickly raced back to my location once they did, and I clambered on board. While it all worked out, I still think about that decision to get out onto a steep cliff. Caution is the best path at all times.

After four decades, field work is still the biggest source of inspiration for my research. Field work has allowed me insights into the life of a large carnivore that lives far from most people and plays a crucial role in its ecosystem, as an apex predator. Polar bears evolved to walk on frozen water where they have a singular focus to survive by killing seals and consuming their energy rich blubber. Being where they live is a privilege. Aside from being personally rewarding, field work has provided insights into the impacts of humans on this charismatic species. Without fieldwork, we would have precious little detailed knowledge of the threats they face.

Acknowledgements

After over 40 years of field research, the list of people to thank is incredibly long. Research is rarely a solo event, the path of discovery requires a team and I thank all those who have helped along the way. I have been incredibly fortunate to work with dozens of excellent collaborators, graduate students, undergraduates and postdoctoral fellows. Polar bear research is heavily dependent on helicopters and I have to provide a nod to all the pilots I have worked with. During field expeditions in the Arctic, I've worked with many Inuit hunters who have shared their knowledge of polar bears and their environment. Their insights have helped shape my research. My wife Kathy made my research possible and I can't thank her enough. Our children, Angus and Emma, have provided motivation to ensure that polar bears persist into the future.

14
Clever Aotearoa New Zealand birds

Isabel Castro

ew experiences are as wonderful as spending enough time with a
species to begin to know some of its secrets, observe moments that
no other human has seen, and understand their world. I have been
a field biologist for about 35 years and have spent at least a third of that
time in wild areas observing, in detail, the life of some amazing creatures,
especially birds but also snails, frogs and mammals. Field work is
addictive – like air, once you've tried it you can't live without it. During my
field work, I've had encounters with wildlife that changed the way I think
and live. These moments made me feel the magic of nature and the
connections we have with other creatures that show how similar we are.
Here, I share a few of these moments and observations with you.

***

I first felt that magical feeling when working on the Galápagos Islands. I
was a young keen biologist on her first real job after completing a Master's
degree. I considered myself lucky to be working in this famous place,
which helped Charles Darwin develop what is arguably the most
important hypothesis in biology – one that explains the diversity of life
and why species have evolved to be the way they are. I had trained in
wildlife survey and assessment, and much of my job on the islands
included monitoring various animal species to understand how their
abundance changed over time.

On one occasion I was counting flamingos. On arrival at the lagoon,
the browns and greens of the water and vegetation, mixed with the sound
of slurping flamingos, and the heat of the day created a sense of
anticipation. I noticed an area covered with many of the typical mini
volcano-like mud nests built by flamingos and, in one corner, hundreds

of snow-white chicks being minded under the watchful eye of a few peach-coloured adults. The adult birds kept the chicks together by rounding them up while walking slowly and seemingly without purpose, in a way that only birds can do. I forgot about the counts and watched their behaviour intently, hearing the feeding noises and observing their interactions. I could imagine easily that they were children in day-care, with guardians ensuring that they were safe and learning a few lessons. Towards the end of the day, other adults started to arrive in large groups that turned the sky pink. The chicks went crazy, screaming in response to the calls of their parents and running to meet them as the adults landed nearby. Soon the flamingo families were reunited and the creche was gone. I was not the first person to experience this aspect of flamingo life, but it was my first experience of a behaviour so like ours that I could easily compare it to the human world. I knew about the risks of anthropomorphism, but could not help but make the connections between us and the flamingos ... At that moment, my career took a turn and I have since concentrated on studying animal behaviour. I work on two related factors: how the behaviour of animals is modified, prevented or changed by human activities, and how we can use animals' behaviour and behavioural plasticity (the capacity of an animal to modify its behaviour in response to changing environmental conditions) for conservation purposes.

***

After the Galápagos, I moved to Aotearoa New Zealand, where I've worked with several endemic bird species. Like Galápagos birds, Aotearoa New Zealand's bird species are quite unafraid and a trained biologist can observe many of their behaviours in surprising detail. My PhD and postdoctoral work involved a behavioural study of the hihi (pronounced heeh heeh), a small endangered passerine bird whose flexible mating system I described. My observations and subsequent experiments helped to create conservation management methods for the species that are still in use today. Male hihi, like males of many other bird species, are showy. Their black heads are topped by white ear tufts and their chests are lined by bright sunflower-yellow feathers. They are also noisy and bold.

Females, on the other hand, are grey–brown in colour and comparatively quiet. Their mating system is a promiscuous one and during the mating season activity is pronounced, with many males actively pursuing the same females and the females either deflecting or accepting their approaches. Perhaps one of the most interesting characteristics of these small birds, which I never studied explicitly but did experience, is their intelligence. I was repeatedly amazed by it.

For my postdoctoral studies, I worked on a small island (135 hectares) where a population of hihi had been translocated. Each bird had a unique combination of colour rings on its legs that allowed us to individually identify them. I used the opportunity to look at the effect of resource distribution on the hihi's choice of mating system and on the time and effort they put into caring for their chicks (parental investment). The experimental set-up involved providing nesting boxes and food, either together or separately, and watching how these extra resources influenced the mating system, parental investment and chick development of the local hihi population. Every morning we'd make fresh nectar mix and provide it in hummingbird feeders – either to the nest site, placing it close to their nesting box, or at a central feeding site. One day, my field assistant and I were sitting at our camp eating lunch when one of our female hihi, whose nest was located about 200 metres away in the bush, flew up to us and started her alarm call. It was a truly weird situation – we were camped in an isolated area, away from the bush where hihi were nesting, and had not seen hihi visiting the camp previously. Nesting females rarely moved more than 50 metres from their nest, so it was unusual for this female to appear so far away from hers. The hihi continued to produce a warning call, a 'tich' sound that gives them their English name 'stitchbird', while looking at us, and then moving to the edge of our camp. She repeated this behaviour a few times. I told my assistant that she looked like a dog trying to tell me to follow. My assistant asked, 'Why don't we?' We got up to follow her. Incredibly, she led us to her nest site where the feeder was out of food. Another nectar-feeding species, the tui (a much larger bird), had tipped the food out of the feeder while trying to access it. We had never experienced this type of behaviour. Maybe the hihi was telling us she needed more food!

After righting the feeder and replacing the nectar mix, we worked on finding a way to stop the tui from accessing the feeder. It was not simple to get the tui to stop harassing the female hihi whenever she tried to visit the feeder. The process also told us more about the intellect of hihi. We trialled caging the feeder so that only hihi could enter and access the nectar. Interestingly, although the tui could no longer get to the food he continued to defend the feeder, preventing the female hihi from entering it. We decided to do something bold with the few options available on a remote island away from stores. We used a cardboard box to hide the entire feeder, making it accessible from underneath only, so that visually the feeder had disappeared. The tui left once it was no longer able to see the feeder, and the hihi soon worked out how to access the food from below. During these procedures, the female hihi kept up her nest incubation routine, apparently tolerating our presence and observing us. She even jumped onto the feeder while we were holding it and grabbed a feed before returning to her job.

I observed numerous other hihi behaviours that I couldn't study further at that time. One area of study that I considered interesting, but which was methodologically challenging, was the behaviour of chicks. I found that hihi chicks, similarly to flamingos, form creches. Male hihi seem to be the babysitters. The creche groups can get very large, with up to 18 chicks in a creche observed during our study. Describing the interactions between the chicks in the creche, and labelling their behaviours (e.g. playing, foraging, vigilance or copying adults), was difficult. Young hihi behaviours often reminded me of the sort of thing mammals do, but wild passerines (perching birds) have not previously been described as doing things like playing. Sometimes I wondered whether I was 'seeing' things that were not there. As my postdoctoral term was ending, I made the choice to simply observe hihi creches without collecting quantitative data. My goals were to gain a deeper understanding of what behaviours occur in these creches and to develop methods for accurately and systematically describing these observations. I headed into the creche area equipped only with my binoculars and a tape recorder. The chicks, all banded with colour bands for individual identification, frolicked in the canopy. I eventually lay down on the forest floor to follow their movements and interactions more easily. That is when I had another of those magic moments. It was late summer, so other bird species were also rearing chicks. Having evolved in the absence of mammalian predators, Aotearoa New Zealand's bird species, like Galápagos birds, are largely unconcerned by humans. Peering through my binoculars, I watched as hihi and North Island toutouwai chicks approached me. At one point, the long-legged toutouwai chicks landed on my folded knees only to be chased away by hihi chicks. It seemed to me that the hihi chicks were telling the toutouwai chicks 'This woman is ours, not yours!' This moment remains vivid in my mind to this day – when my study bird 'defended' me from another species. Perhaps the hihi chicks viewed me as a source of resources, since I had been providing them with supplementary food and nest sites. We also measured the chicks regularly, from hatching to fledging, and I wondered if they had imprinted on us. Whatever the reason, how many people get to experience that kind of moment? This experience increased my awareness of the effect that humans can have on other species and how what we do to them, good or bad, can modify their behaviour. If their behaviour changes because of something we've done, how does that affect our inferences and what are the implications for conservation?

***

Hihi were not the only smart birds living on the island with a capacity to solve problems in their everyday lives. The feeders we used for our experiments with hihi consisted of traditional hummingbird plastic feeders with a transparent bottle screwed to a red base. Several small yellow flowers with a central opening were attached to the base, allowing access to the nectar within the well of the base. North Island tieke, another Aotearoa New Zealand endemic bird species, were very curious about our research activities, often being present when we established the boxes and feeders for our research with hihi around the island and when we changed feeders and/or collected observations of hihi behaviour. Several weeks into our experiment, we noticed that the yellow flowers had often been detached from the feeders and left lying on the ground. We had no idea how it was happening. We had not observed any other species using the feeders. We decided to place a camera near one of the feeders to capture the culprit in action. To our surprise, it was the tieke. This species, better known for the cultural transmission of their song, is a medium-sized passerine with complex social behaviours not yet fully described. They are fiercely territorial and produce strident calls of one sound (for females) or a few sounds (for males) repeated over and over that can be heard a long way away. Each male call is unique to that bird, and all the birds in an area have their own type of sounds that are used to compose the individual's calls, or dialects. Birds that live on the edge between two areas learn the sounds of both areas in a process that we call cultural transmission because the young birds copy the sounds of the adults nearby when they are learning and developing their own songs. Some tieke sounds are also copied from other species, including human-made sounds, especially loud sounds such as ambulance sirens. Their foraging behaviour is extraordinary because they use many techniques to get food – from digging to gleaning, probing, and opening leaves to extract hidden insects. They use their bill and legs in similar ways to parrots. For example, tieke hold branches or leaves with their feet and use their bill to reach the food in or on them – a very unusual behaviour for a passerine. They are omnivores, so their tongues have feathery endings which could be useful when extracting nectar, but their tongues are not long and thin like those of specialised nectar-feeders. Rather, their tongues are thick and wide. In our video, the tieke approached the feeder and tried several yellow flowers with its tongue. To our amazement, after these failed attempts to suck nectar, the bird grabbed the edge of a yellow flower and flicked it sideways, making it fly to the ground, then proceeded to suck out nectar with its tongue from the now wider hole! This occurrence was not a one-off, as flowers were missing from several feeders around the island. We don't know whether tieke passed on the technique to one another or if they each independently developed the technique.

It was clear that tieke can solve problems, and benefit from their ability to do so. Animals have intellects that allow them to solve the problems they encounter in their environments. However, we seldom see these behaviours for what they are (problem-solving) because behaviours of wild animals do not always immediately appear to be a response to a problem. If the hihi feeders were a natural source of food to tieke (and not a human-provided item), I would have most likely not been amazed by seeing the tieke flicking the yellow flowers away. I would have thought that tieke just used the resource in that way. In fact, I may never have found out that they were flicking the yellow flowers away at all, as it is likely that I would have found the feeders without them, missing the beginning of the innovation.

***

Intermediate school in Aotearoa New Zealand must be the most fun school phase of all time. Children of 11–12 years of age are encouraged to explore arts and science through projects that they plan and carry out. My 11-year-old adoptive niece, Clea, wanted to do a science project for her intermediate school. I have a long-term brown kiwi project on an island off the east coast of Aotearoa New Zealand, where we have marked individual kiwi so that we can follow their whereabouts over time. Kiwi are relics of a long gone period in avian time and evolved without the presence of mammalian predators, so they have characteristics and behaviours that are rare in modern birds. In fact, kiwi species occupy the niche of a nocturnal ground insectivore, which is a role now filled by mammal species in other parts of the world. Kiwi are adapted to life in the dark, where smell and sound are more useful than vision in navigating the challenges of life. For example, kiwi have an organ at the tip of their beak (the bill-tip organ) that allows them to sense vibrations from invertebrates underground (a sense called remote touch). Their brain, similar in size to that of parrots, is larger in areas that are used for the integration of information coming from their senses. Kiwi perform many odour-sensing behaviours such as sniffing (unusual among birds), which allow them to acquire and interpret the world in the dark.

I was about to start a pilot trial for an experiment with kiwi. A couple of years earlier, with one of my PhD students Susan Cunningham, we learned that kiwi could find food using the sense of smell alone or in combination with remote sense. This time I wanted to know whether kiwi used the sense of smell in other settings. I thought Clea would enjoy being involved in it for her school project, and I would of course help and accompany her in the field as most of the work would be undertaken at night. Clea had been one of my field assistants since she was seven, so she was already experienced. When I told her about the experiment, she thought it sounded like the best project ever. The project involved presenting kiwi with samples of items in their local environment and videoing their response, for later analysis. We presented kiwi with banana skins to see whether they would be interested in a novel scent in their path; kiwi faeces, as an example of a social scent; and sheep faeces, in case kiwi had an interest in faeces more generally. We also collected data on kiwi behaviour when there were none of these items in their path, to make sure that what they did when one of our items was in their path was specific to the item. We spent many nights heading into the field as night came to turn on the videos and then heading home after videoing the birds, around midnight. Usually, we saw lots of kiwi running away from us as we made noise on the tracks, but not always ...

On one occasion, we were emerging from the bush onto a track that followed a farm meadow when both of us heard the loud, and by now easily recognisable, noise of a kiwi walking through the undergrowth. We froze, turning our heads (and head torches) in the direction of the noise, which was now very close to us. The kiwi also stopped and looked in our direction. It was a medium-sized bird, perhaps a male or a juvenile female. Unfazed by our light it 'sniffed' us loudly, then walked decidedly towards me. On reaching my left foot it raised its head and pecked at my boot with all its might ... Then it shook its head (perhaps hurt or just surprised by the unyielding material) and walked away into the night in the characteristic kiwi way, by tapping the ground ahead with every step. Clea and I looked at each other and giggled. What an extraordinary encounter with a species that most people would never see because of its nocturnal habit, but sadly also because it is now very rare. Another night, as we sat in the dark waiting for the kiwi to be videoed, one of the males (Dario) approached us. When Dario was very close to us (we didn't have head torches this time) he sniffed us, waited, then sniffed us again. This time I could not resist returning the behaviour. I imitated his sniff, doing my best impression of a kiwi. He was surprised and ran a few metres away from us but then stopped, sniffed again, and continued to move away. We giggled again and have done so every time we remember the story. By the way, we found that kiwi were most interested in other kiwi faeces, then in banana skins and lastly in sheep faeces. We interpreted this as meaning that they use their sense of smell both in social contexts and to explore their environment.

***

These observations were opportunistic and my telling of them is anecdotal. They were not collected to test a hypothesis, so they are not publishable in a scientific sense. However, informal observations made in the field are the building blocks of the ecological sciences, and often become disproportionately important to the conservation of a species with time and testing. They speak of other species – non-human species – having a much richer life and intellect than we tend to ascribe to them. This is especially true for wild birds. For a long time, birds were considered less mentally capable than mammals. Recent scientific studies of birds' brain function and behaviour, following observations such as the ones I've shared with you here, are showing that their cognitive capabilities are often greater than those of similar-sized mammals. I have always loved animals and respected them, but these field experiences have increased my awareness. I consider my study species' intelligence much more these days when I design experiments to learn about or decipher their behaviour. I hope that these anecdotes inspire other researchers, and perhaps encourage them to design their own research projects influenced by some of these observations.

Acknowledgements

I am deeply indebted to the many people who made my research possible. First whānau, hapū and Iwi who gave me permission to work in their rohe (lands): Ngati Toa Rangatira kaitiaki (guardians) of Kapiti Island; the Te Arawa people of Rotorua who allowed me to live on Mokoia Island while doing my postdoc, in particular Mita Muhi, who introduced me to Mātauranga Māori and shared the taiaha camps with me and my team, and John Marsh who taught me so much about the Māori culture and how to approach knowledge. Ngai Tai have supported my research with kiwi in their rohe for 20 years and counting. Many thanks to my whānau in Ipipiri (the Bay of Islands) Ngati Kuta and Te Patukeha hapū for supporting, accompanying and helping me wherever I go in Aotearoa, especially Richard Witehira and the 3B2 Ahu Whenua Trust. The Chamberlin family have allowed me to roam their land with my students and volunteers to study kiwi for over two decades, and have participated in the field work. I am indebted to my postgrad students who worked hard in the field and contributed to my own research as well as doing their own. I am thankful to my collaborators, especially Maurice Alley, Dianne Brunton, Stephen Marsland and Peter Lockhart, whose friendship and company in the field have made it all so much fun. Funding came from many sources, primarily the Royal Society of New Zealand (Marsden Fund), the World Wide Fund for Nature, the New Zealand Ministry of Business Innovation and Employment, Massey University, Te Punaha Matatini, Save the Kiwi Trust, and the Birds NZ Fund. Thanks to literally hundreds of volunteers who have contributed to my projects since I started working in Aotearoa New Zealand.

15

Expect the unexpected when dealing with the devil

David G. Hamilton

Night in north-west Tasmania. The sky is overcast and not even the moon peeks through the thick clouds. Perfect devil conditions. We're sitting in a utility vehicle (ute), parked about 10 metres from the staked-down carcass of a pademelon, ready for action. We are here to observe Tasmanian devils and are listening intently to a small radio for signs of their presence, but after sitting in the cold and pitch black for over two hours, we haven't heard or seen a thing ...

This tried-and-tested method of observing devils is a relatively simple set-up, although it has required multiple, often bizarre, conversations to get to its current state. The lure, a recently deceased pademelon, was one of around 50 enthusiastically delivered to our door the previous night, following a slight misunderstanding with a local about what constituted 'a couple'. The repurposed technology (essential to any ecological endeavour), a baby monitor, is set up next to the pungent carcass. Imagine, if you will, the look on the face of a salesperson when you tentatively ask them if a baby monitor is weather-proof. When the unmistakable noise of devils scuffling is heard on the car end of the baby monitor, the red lights rigged up around the carcass are slowly turned up and observations can begin!

That's the theory, anyway. However, we've been sitting in the ute, waiting for the baby monitor to go off, for almost four hours. We're becoming cold and disillusioned, resorting to slowly turning the red lights up every half-hour or so just in case a particularly crafty devil has snuck past us.

Thirty minutes ticks over. As the lights are dialling up, we notice something slightly different about the pademelon carcass. Has it moved?

No, but it somehow looks smaller than before ... Oh dear. One of its legs has disappeared.

This method may require a rethink ...

***

A particular idiom often pops into my head while I'm conducting field work – 'better the devil you know than the devil you don't'. Admittedly, I regularly work with Tasmanian devils and my mind tends to run on puns, but it also seems an appropriate phrase to apply to various scientific endeavours. Scientists study plants and animals because we seek to understand them better – that heightened understanding often helps us manage novel or challenging situations, whether they be disease outbreaks or invasive species. However, sometimes understanding the devil you know is only half the battle.

I learnt this particular lesson while conducting my PhD field work. Devils are large carnivorous marsupials that are threatened throughout their range in lutruwita, Tasmania. My research focused on understanding the ways that devils interact with one another, and how these interactions were driving the spread of a novel disease across the devils' island home. Devil facial tumour disease (DFTD) is a transmissible cancer, an incredibly rare form of disease about which we still understand terrifyingly little. While cancers are normally restricted to the host bodies they originate in, DFTD can jump from devil to devil when they bite one another (which they do frequently during feeding and mating), causing wounds in and around the head that cancer cells can enter. Consequently, the cancer's spread is inherently linked to the behaviour of the animals themselves, which is what I was to set about investigating.

The first phase of my project involved studying a population of devils as yet unaffected by DFTD. I aimed to look at how the devils interacted with one another in the absence of disease. In the absence of DFTD, I could track factors like bite wound occurrence over time without worrying about confounding disease-related variables. In Tasmania, getting away from the extent of DFTD meant travelling to the very north-west of the state to find a disease-free population. Tasmania's Wild West is so-called for various reasons, ranging from its climate to its inhabitants, so it served as quite an introduction to the state.

In fact, my formal introduction to Tasmanian devils themselves had come only the week before I started field work. I was testing technology that was critical to my research – proximity logging collars that were to be fitted to individual devils. Each collar emitted a unique ultra-high frequency (UHF) pulse that could be picked up by other collars, thus logging when two collar-wearing devils came within a pre-determined distance of one another. I set this distance at 30 centimetres – the distance at which two devils could conceivably bite one another (and thus, potentially transmit DFTD). However, the technology was fairly new. The last thing I wanted was to deploy the collars on wild devils and then discover that there was an unexpected problem, so the plan was to trial them first on a group of captive devils housed at a local wildlife sanctuary. Fitting collars to the captive devils was not straightforward, with most behaving like the whirling dervishes portrayed by certain cartoon Tasmanian devils. Fortunately I managed, and the technology worked like a charm. While my mind was at ease about technology-related issues, I was now worried about the logistics of fitting the collars on wild devils! If captive devils were this difficult to deal with, I suspected that their wild relatives would give me hell.

In fact, this is not where the lesson arrives, and nothing could be further from the truth – wild Tasmanian devils are a dream study species in several ways, one of which is the response of 99% of individuals to the novel event of being captured and handled. Devils are the alpha carnivores in the Tasmanian landscape, so their response to being handled by what they must perceive to be a predator is (somewhat unexpectedly for an animal that has one of the highest pound-for-pound bite forces of any animal species on the planet) just to freeze and hope it goes away. This response makes it surprisingly easy to take measurements, check their face and mouth for signs of disease, and fit collars. To this day, every person I take into the field for devil monitoring is astounded by how placid they are during handling. People expect chaos, mayhem and a possible hospital visit. Some seem almost disappointed to find that wild devils are more like teddy bears when handled. I was once accompanied by a badger researcher from the UK. Badgers are a similarly sized carnivore, with a polar-opposite temperament. Trapped badgers have to be sedated from a distance before handling, using a syringe on the end of a long pole, or the handler may lose a finger. The researcher's response on seeing a devil handled for the first time was a kind of jealous confusion, with a few curses and mutterings on how easy wildlife researchers had it in Australia.

The ease of devil handling made fitting collars to the entire adult population of my north-western study site relatively easy. Every devil I targeted was caught and collared within the first month, and I still had all my fingers! At this point, I was to learn one of the first major life lessons of my PhD – however prepared you are, you can never be prepared for everything.

I was around two months into my field work, with no major hiccups. One of my biggest worries had been about the fit of the collars – devils are kind of sausage-shaped, so their necks are not obvious. This sausage-like tendency meant it was critical to check that the collar fitted perfectly on every animal. An ill-fitting collar might be too tight and start to rub, or be too loose and slip off. I was delighted that I'd got fitting the collars down to a fine art, with most animals turning up regularly with perfectly fitted collars. It was not to last, of course. One otherwise fine day in autumn, my luck ran out. A previously collared animal turned up naked-necked, his collar missing-in-action. Worse, it was one of my most reliably caught devils – a young male who, until that point, had turned up like clockwork for his collar-checks. In fact, I had received some fantastic insights as a result of data collected from his collar. It was the middle of devil mating season, and this particular male had displayed some fascinating behaviour. During the mating season, male devils will mate-guard females for up to two weeks at a time – mating regularly and trying to ensure that other males cannot. One hulking 14 kilogram monster of a male devil, face worn with battle scars, had mate-guarded around half of the female population at various times during the season. Every time he finished with a particular female, this younger male would immediately scurry in and spend a few hours with her – presumably sneaking in a mating session while the big devil's attention was elsewhere. Tasmanian devil litters can have mixed parentage, which is likely influenced by their approach to mating, so it was interesting to see different tactics in action.

Anyway, this young male – my most reliable provider of fascinating data – was the first devil to ditch his collar. I was devastated. If this devil, who I was certain I'd done everything right with, could turn up collarless, surely it was only a matter of time before the rest of the devils started losing their expensive bling.

I needn't have worried. Each collar also had a very high frequency (VHF) component, which meant it could be tracked remotely. Once I had finished devil-trapping for the day, I set off in search of the young male's collar. I had tracked him to his preferred denning spots regularly, so already had a good idea of where he liked to frequent. Finding a signal for his collar didn't take long. The signal was strong – a good sign, as it suggested that he hadn't dropped the collar underground, where it would be almost impossible to retrieve. I triangulated the signal to the area around a small bush and was about to dive in to look for the collar when I spotted something strange at my feet. A bandicoot-sized hole perfectly lined with some shiny black material. It was the missing collar, wedged halfway down the hole. The unwitting young male had jammed his head into a space where it didn't fit – when he managed to yank it out the collar was left behind, a perfect fit in the snug hole. I started laughing uncontrollably, out of relief as much as anything. You can prepare for many things – known seasonal changes in devil body mass or the aggressive behaviour of the animals themselves, for example. However, unexpected behaviours of individuals (somewhat ironically, the very thing I was trying to study) can throw plans into disarray. Basically, I could not have prepared for the possibility of a devil randomly sticking his head down a tight-fitting bandicoot hole, which, no doubt aided by the devil's sausage-like anatomy, was enough to make the collar slide off. Somehow, understanding this reality can simultaneously be both a comfort and a concern.

The following few months passed without major incident. A few more devils briefly lost their collars and had them refitted, but most were remarkably well behaved – bandicoot-chasing season must have been over. I had planned for the collars to remain on the devils for six months, which encompassed the battery life time and would produce data from both mating and non-mating periods. Once the six months were up, it was time to start removing collars. For about two-thirds of the devil population, this task was relatively easy. The devils had turned up regularly during trapping sessions for the entire study period and weren't about to change their ways at the close. The hold-outs were always going to be the trickier customers. I generally had to target these devils with traps placed near their dens; a couple were particularly stubborn, evading capture for some time. After around six weeks of intense effort, I was down to the last two devils – a pair of adult females who were hunkered down at opposite ends of the study area. They both had favourite denning spots, so I was able to find them consistently. It was catching them that was proving difficult. The first female seemed to have taken up residence in the middle of the densest population of spotted-tailed quolls (a spotted relative of the Tasmanian devil that is lured into traps by the same bait that attracts their larger cousins) I have ever encountered. I had lugged multiple pipe traps 3 kilometres along the beach to place them around her den, and every morning I was met with the same result – an army of captured quolls and not a devil in sight. Ultimately, I caught her. Not in one of the traps that I had painstakingly dragged along the beach, but in one of a handful I kept set across the study area 'just in case'.

The final devil was not in such a helpful mood. Eventually I caught her in the most unexpected of circumstances, and again, it was something I could not have prepared for. She had been frequenting the same den for months, a lovely spot in the dunes in a pre-excavated wombat burrow. I located her there every day for weeks, until one day she wasn't there. Panicked, I attempted to find her signal by combing the whole area, checking spots I'd never suspected she might use – nothing, not a peep. Late in the day I checked one final spot and was rewarded with a strong signal. She was close! I was near a nice patch of open eucalypt woodland, and she seemed to be in it somewhere. I triangulated her position in the woodland and walked around looking for a suitable denning spot. There wasn't much around – it was hard ground with little burrowing opportunity, not even woody debris that could be tunnelled into or under. I was flummoxed. The signal was stronger than ever, but I couldn't see an obvious shelter. Exasperated, I bent down and started scanning the ground, looking for diggings, scat, anything. I turned towards a patch of bracken on my left and suddenly I was face-to-face with a Tasmanian devil. The final animal. Staring at me intently, not from a nice safe underground den but right there in the open. Thoughts raced through my head – I was totally unprepared for this kind of encounter. I couldn't very well crash tackle her and just take the collar off. I had no equipment with me besides my tracking gear. Everything I needed was back at the car – only around 200 metres away, but it felt like a million miles. Slowly I backed away, trying not to frighten her. Once I'd left a suitable distance between us, I bolted the rest of the way to the vehicle and quickly grabbed a hessian sack (standard devil-containing equipment) and all the tools I needed to remove the collar. I snuck back to the patch of bracken. By some minor miracle the devil had remained in the same spot. Now for the fun part – catching a devil by hand. My heart was pumping fast; this chance might be my only one. What if she gets away, what if she bites me!? Slowly I moved my hand around behind her. She watched me the whole time but didn't move, until suddenly I had her firmly by the base of her nice thick tail. She started kicking and screaming, but I somehow managed to slide her into the hessian sack. Then she went into standard devil-being-handled mode and quietened. Terrified that she might pull a swift one on me and decide to make her exit, I rapidly removed the collar. Then I gave her a full health check and took a couple of measurements. She was in fantastic condition with no sign that the collar had been too tight. I was relieved, euphoric even, as I released her. Second major lesson learnt, and it countered my first. Always prepare as much as you can, particularly when you're working with wild animals who are also individuals! Prepare for every outcome that you can think of, and hopefully you'll be ready as you can be when unexpected situations or opportunities arise. In this case, I was able to make it work, and watching that devil running off into the bush remains one of my favourite memories in field work. Better the devil you know, indeed.

Acknowledgements

The field work outlined here was conducted on the lands of the peerapper people of the takayna region. For thousands of years they have cared for and protected the lands, waters, plants and animals of this wonderful part of lutruwita, Tasmania, and I pay my respects to their Elders past and present.

16

Outfoxed by a jackal, and other tales from the Indian savanna

Abi T. Vanak

I was lost in the Golan Heights.

I consider myself quite adept at understanding accents, but clearly something had been lost in translation. Did Alon from the National Park Service say that I was supposed to turn left on the way to the Beit Kefet Forest or at the entrance of the Beit Kefet Reserve? Was I even supposed to be here, in this extremely militarised zone on the Israel–Syria border? I was fairly certain that the pimply teenagers masquerading as soldiers, their Uzi submachine guns casually slung over their shoulders, would not take kindly to a random foreigner driving into a military compound. Of course, that was precisely what I did. It took some explaining to get out of the situation. The magic words were 'I'm from India'. Suddenly, the heavily armed soldiers became friendly. India is one of the prime destinations that Israeli soldiers visit to unwind after their mandatory army service. After several *slichas* and *todas* I was pointed in the right direction.

Eventually, at the end of a forest road bordering an agricultural field, I found Aviv. He was annoyed at my tardiness. I was supposed to have met him half an hour ago. I apologised profusely for my poor sense of direction and not knowing the subtle differences between 'take this left turn' and 'the other left turn' in Alon's heavily accented English. However, I was quickly distracted from my apology. Aviv was standing next to a red fox. The fox was not happy.

Aviv is a jovial, stocky middle-aged man, who would have not been out of place behind the desk of a reputable law firm. However, here he was – in the Golan Heights – standing next to a trapped fox. Aviv is a licensed trapper, and I was here to learn how to trap golden jackals from him. With remarkable dexterity, Aviv released the fox from the leg-hold trap. He caught it by the scruff of its neck with one hand, took its hind legs in the other, and gently let it go. The fox looked dazed for a moment, but once it realised it was free it bolted off into the scrub, clearly uninjured by the trap.

We moved on to check the other traps. The next three had not been triggered. However, in the fourth trap, next to an abandoned stone barn, a young jackal was fast asleep, right foot held firmly in the grip of the trap. Aviv expertly released the jackal from the trap using an animal restraint snare, and because the jackal was too young to process (measure) it was let go. He proceeded to show me how to set up traps, what lures to use, how to apply them and, most importantly, how to use the trapper's version of smoke and mirrors to trick the wily jackals!

***

You know what they say about clever foxes. Well, as it turns out, jackals are cleverer, and catching them is tricky. For ecologists interested in how animals move and use their environment, it is standard practice to catch an animal, attach a tracking device and let them reveal their deepest secrets. It sounds simple. In theory.

As hunter-gatherers, early humans invented various ways of catching animals for the pot. Since the point of capturing an animal was ultimately to kill it, reducing its stress and suffering was not a high priority. However, as biologists, we don't want to hurt study animals or cause them more stress than is necessary. Consequently, we ensure that our catching of the animals is safe, humane and involves the least possible amount of stress and discomfort.

Of course, most animals don't want to be subjected to this ignominy and will go to great lengths to avoid getting caught. I first learnt to trap foxes with Dr Brian Cypher from the Endangered Species Recovery Program at California State University, Stanislaus. Brian worked with the endangered San Joaquin kit fox that was very similar to the Indian fox that I had planned to study as part of my PhD research. They trapped kit foxes using cage traps left overnight in the streets of Bakersfield, California. On my very first morning checking traps, we had caught five foxes. I was delighted. If I had even a tenth of this success rate, I would have no trouble answering my research questions. 'Just throw some hot dogs in there. Foxes will walk into anything for a wiener' explained Brian nonchalantly.

I returned to India a few months later, armed with this knowledge and a couple of bottles of very foul-smelling lure, hoping to smuggle it past Customs in India. My field site was in the savanna grasslands of Maharashtra in western India, near a town called Solapur. It was to be my home for the next two and half years. My first challenge was to make traps. Since India does not have a trapping industry, and not many people study wildlife by trapping them first, you can't just pick up a catalogue and mail order Tomahawk traps, for example! I had to get them fabricated. Luckily, I made a few friends in Solapur who were part of a local nature lovers' group, colloquially called *sarpa mitra* or 'friends of snakes'. Their main activity was to rescue snakes from backyards or houses and release them into the wilderness.

Abhijeet Kulkarni was a member of *sarpa mitra*. At that point, he was a farmer and paneer (a type of Indian cottage cheese) maker. He was also a jack of many trades. We brainstormed trap ideas together, and Abhijeet convinced a friend of his, who made commercial freezers, to let us use their factory floor for making the cage traps. We followed a quintessential Indian philosophy – *jugaad*! To those familiar with the 1980s American television show 'MacGyver', this phrase would be the equivalent of 'MacGyvering' – to be inventive, innovative and frugal. After several days and much clanging, bending, welding and breaking, we had our *jugaad* cages ready for field testing. Suffice to say, these prototypes were not going to win any awards for design or elegance.

However, and perhaps astonishingly, they worked! We caught domestic cats, a jungle cat and even a couple of babbler birds, but no fox. The foxes were there, they came to the traps and ate the bait at the entrance of the trap, but no amount of tuna or Solapur's famous grapes would lure them inside the cage. Abhijeet, who by now was working full-time with me (he sold his buffalos and his paneer business), and I tried every trick in the book. We covered the traps with grass, we half-buried them in bushes, we sprayed them with lure, we changed design so that they appeared like a tunnel rather than a human-made square. But zilch, *ille, nada*, no luck. We were clearly the only ones complaining about sour grapes.

After a few weeks of futile effort, we realised that we needed to change tack. My PhD advisor, Professor Matt Gompper, suggested using padded leg-hold traps. I was apprehensive. Would they harm the foxes? After talking to several researchers who had successfully used them for other species, my fears were somewhat eased. I ordered a set of rubber padded foot-hold traps from the US, and my team and I tried to figure out the best way of deploying them. Unlike my American counterparts, I could not set the traps overnight, leave them and check them in the morning. My study site was in a heavily human-dominated landscape, and free-ranging dogs were a danger to trapped animals. I had to figure out a way of installing an alert system on the traps that would be triggered upon activation.

I remembered that a colleague had set up an ingenious system that involved rigging a radio-collar to his traps, which would turn on if the trap was activated. The radio-collars had a magnetic switch. When the magnet was removed the collar turned on. Monitoring the collar frequency at a regular interval allowed the researcher to determine whether an animal had triggered the trap and likely been caught.

Our system in place, the traps set and baited with lots of grapes (yes, it's true – foxes like grapes), we waited. Not long after, we heard a 'beep-beep-beep' on the receiver. Trap number two had been activated. We rushed to the trap site – sure enough, our very first Indian fox was in front of us. She was in a panicked state, trying to escape the metal beast that had sprung up out of the ground and grabbed her. We quickly restrained her using a net, and covered her with a cloth bag so that she would calm down. Wildlife veterinarian extraordinaire, Dr Aniruddha Belsare, administered a sedative that would allow us to safely handle the fox and reduce her stress. However, there was a catch. The dosage rate was usually calculated based on the animal's body mass. Since she was the first fox that we had caught, we had not been able to calibrate our visual estimation of the animal's estimated body mass to its actual mass. We heavily underestimated her mass, and the sedative wore off before we could complete all of the measurements we needed to take. Aniruddha very calmly caught the fox by the scruff of her neck and told us to go ahead and finish quickly. Measurements made, animal weighed, collar fitted, magnet removed, and we were done. Off went NNJ-Fl, or Misscal as we named her soon after (because we miscalculated her dose).

Over time, we perfected our routine and catching foxes became child's play. During my study, I trapped over 40 foxes, the odd jungle cat, a mongoose, several dogs and a wolf that did not have the patience to wait for us, and simply yanked its leg out of the puny fox-sized traps.

So why did I need to go to the Golan Heights to learn from an Israeli trapper?

I had returned to India after finishing my PhD in the US and a three-year postdoctoral position in South Africa. I was now a faculty member, and an independent scientist. I was supposed to guide students and be the Grand Poo-bah who knew everything. I had just secured a large grant to study a guild (a group of species that have similar requirements and play a similar role in a community) of mesocarnivores in human-dominated landscapes near Baramati town, not very far from where I caught my first foxes. This time, I would be using GPS collars instead of the older VHF radio-collars to track the animals. The movement data would be more precise, more frequent and far less likely to suffer from the biases of manual tracking associated with VHF studies.

I was excited because not only would I be catching Indian foxes again, but also the co-inhabitants of India's savanna grasslands and agricultural landscapes – the jungle cat and the golden jackal.

We set up our trapping systems again. Abhijeet was back on the team, and we had new students and research associates to train. We were a little rusty to begin with but soon fell into our old groove, and the foxes and jungle cats came rolling in. However, several weeks in we still had not managed to catch a single jackal. Assuming that the jackals would be just as easy to catch as the foxes, we had used the same system. Set the traps, use chicken offal as bait, place a drop of lure on a stone next to the traps, and attach the trap trigger to a string.

We waited and waited. Something weird was going on. The bait was gone or the traps had been dug up or the string had been pulled, setting off the alarm – but there was never a jackal. It was almost as if they *knew* where the traps were. We decided to install a camera trap at our trap sites to see what was happening. Sure enough, we caught the culprits stealing the bait, without getting trapped!

The jackals clearly realised that something suspicious was going on. They would approach the bait then, astonishingly, sit down and creep forwards so as not trigger the traps. The traps were buried in the ground, and an animal must put its foot on the treadle to activate the padded jaws. Somehow, the jackals had found a way to avoid setting them off.

One male, who we later named Don after the iconic main character in a Bollywood movie of the same name ('It's not just difficult to catch Don, it's impossible!!'), figured out that he could dig out the chains that anchored the traps, drag them out of the ground and then happily skip around them to enjoy a sumptuous feast of chicken feet. We couldn't figure out what we were doing wrong and I needed expert advice, which is why I found myself wandering into army camps on the Israel–Syria border looking for Aviv!

'You can't get your smell on the traps' chided Aviv. 'You have to first treat them.' Eh? I had never needed to treat the traps for foxes! Well, apparently jackals are more wary than foxes. According to Aviv, you had to wash the traps meticulously, dip them in wax, then bury them in the soil that you intend to eventually set them in. The wax masked the smell of metal and absorbed the smell of the soil. Gloves had to be worn when handling the traps, and you had to stand on a mat when setting them so that your sweat and smell weren't left on the ground. Then, it was important to grab a few leaves of the nearest aromatic plant and brush everything with it to mask whatever remaining smell you'd left behind. Also, bait could not be placed on the ground. The jackal should not approach the trap with its nose on the ground. It should instead sniff 'up' distractedly. Putting the bait on a tree or bush encouraged them to do that. A secondary trap site should also be prepared, but without bait. A drop of lure had to be placed away from the main trap set on a rock, and the traps set around the rock. The jackal would come to investigate this strange odour and then, like all good dogs, turn around to mark it, and bam!

Armed with these tricks, I rushed back to Baramati and attempted to fool the jackals. The process worked like a charm. We were catching jackals at the main trap set, at the peripheral ones, on the paths and at the scent-marking rock. We were elated. Finally, the data started coming in, and we learnt that the foxes used only the grasslands, the jackals mainly used the sugarcane fields and the jungle cats used everything. One large male jungle cat, named Sultan, even walked the streets of Baramati, presumably bullying the domestic cats in the alleyways.

The jackals and jungle cats benefited from their proximity to humans. They had plenty of food, and the sugarcane fields provided excellent shelter. The poultry farms disposed of dead chickens carelessly along the roadside, which provided a feast for all the animals. However, this bounty came at a cost to the mesocarnivores. One day, we found an entire pack of jackals dead. We suspected that they had been poisoned. Don, Roma and Jet Li all perished. In another bizarre incident, an electricity line felled in a storm left a trail of carcasses along the road. A monitor lizard, a jackal (Langda Tyagi this time) and two jungle cats were all electrocuted in a line. We wondered how this situation happened. Did the dead monitor lizard attract the other predators which were all, one by one, lured to their death?

We knew that as long as suitable habitat remained, these animals would eventually be replaced by other individuals, replenishing the population. The greater danger lay in the rapid changes that were coming to the landscape. A new six-lane highway was under construction. With its completion, more industrial activity would ensue. As land prices increased, the remaining grasslands and fallows would soon get parcelled up as residential layouts and the foxes would be squeezed out of their homes. The jungle cats would likely hold on for a long time, and the jackals would persist if the sugarcane continued to be irrigated through canals.

Catching animals is thrilling. Ask any hunter or fisher. It's the thrill of the chase and all that. However, as I mentioned earlier, scientists are different from hunters. We don't intend any harm to the animal, and we ensure that the entire capture process causes them as little stress as possible. However, no matter what precautions you take, sometimes things go wrong which could result in injury to the animal, or worse, death. The danger is not just to the animal. When large mammals are captured, humans are also at risk. These risks can be mitigated by having a well-qualified team, careful planning, having the right equipment and, of course, experience. Many of us biologists justify our work because of the valuable information that these animals provide. The benefits are huge. We can peek into the secret lives of animals, to discover how they are adapting to the ever-changing landscapes around them. As sensors and tags get smaller and technology continues to improve, some scientists are even attempting to build an 'internet of animals'. Equipped with micro-sensors, these animals can allow us to map the environment almost exactly as they perceive it.

Our study is now one of the largest movement ecology studies in India and it now includes another species, the much larger striped hyaena. Trapping this animal has its own set of challenges, but that story will have to wait for another day.

17

Finding frogs in the most unexpected of places

Jodi J. L. Rowley

I had a simple but difficult mission. To find a tiny mottled green and brown frog that hadn't been seen in over 40 years, which also lived in some of the most rugged terrain in Australia.

The amphibian in question was the peppered tree frog, a small frog just a little bigger than my thumbnail and known only from a handful of rocky streams on the Northern Tablelands of New South Wales, eastern Australia. Since being discovered and named in the 1980s, this tiny frog seemed to have vanished. Several survey expeditions had searched for the species in the 40 or so years since it had disappeared. However, it was like finding the proverbial needle in a haystack. Potential habitat, where the species could be hiding, was expansive – a vast landscape of granite boulders, waterfalls and sheer cliffs. Previous searches had only just scratched the surface. If the frog was still out there somewhere, it would be one of Australia's most threatened frog species and in desperate need of our help.

I began my search for the peppered tree frog in the spring of 2016. Like those who had searched for it previously, I started with the most obvious places – the five spots where it was originally found. These locations were all streams that flowed from the highest parts of the Tablelands, over 1,000 metres in elevation, towards the Pacific Ocean. Most of these streams cascaded steeply off the Tablelands, becoming waterfalls as they flowed east. The locals called the area Eastern Falls country.

My colleagues and I searched these five streams and dozens of others surrounding them. We hiked alongside them as they flowed across the plateau, then climbed down as far as we could. The streams were challenging to survey as they were very rocky, often with enormous granite boulders, sheer cliffs and scree slopes. The boulders that we clung to weren't always particularly stable. Overall, it made for some precarious moments.

While we found frogs, lots of frogs in fact, we didn't find our target, the peppered tree frog.

From the historical peppered tree frog locations, our surveys expanded out and into some of the most remote parts of the Northern Tablelands. We would stay nearby, not far from the town of Glen Innes, spending one to two weeks surveying for frogs every night then returning to Sydney, and travelling back up the next month to search again. We began each day full of hope that we'd find our tiny target. After poring over topographic maps, asking for advice from the local community and modelling local environmental conditions, we'd pick a section of stream that we deemed potentially suitable habitat and within the likely range of the peppered tree frog. We'd leave our base camp just after lunch, driving as close as we could to the chosen site, then hike into the forest to survey frogs along a stream. Once we'd gone as far as we could – determined either by the stream plummeting off a cliff face in front of us, or it simply being too late – we'd hike back to the car and drive home. The surveys were hard work physically, and typically ended in the early hours of the morning.

We repeated our blocks of surveys every month of spring and summer, and into autumn. We were rewarded with sightings of platypus so close that we almost accidentally stepped on them, with lyrebirds displaying right in front of us and with inquisitive whiskered quolls watching us warily. We saw orchid-covered boulders, enormous carpet pythons and a sky full of bright twinkling stars on still nights. We also saw frogs aplenty, but not the peppered tree frog.

The following spring, in October 2017, my colleague Tim and I were again up on the Northern Tablelands, conducting our second season of surveys. On this trip, we had planned for 10 nights of surveys and spent the first few nights re-checking places where the frog was once found – just in case – before searching further afield. We set out to survey one of the most remote parts of Guy Fawkes River National Park, a wild beautiful place on the eastern edge of the New England Tablelands. We were interested in a section of the Sara River named Starlight. This site has been recommended to me by several friends in the Glen Innes community, although each recommendation came with a warning about how challenging it would be to access. Decades ago, a vehicle trail of sorts had meandered through the forest, which ran from the Tablelands down into the valley where Starlight was located. Few dared to drive down, and reportedly those who did had to tie a huge log to their front bumper bar in preparation for the drive back up. The log was to weigh their car down at the front, hopefully preventing it from tipping up and over backwards on the extraordinarily steep trail, and tumbling back down to Starlight.

We left the house we were staying in, just outside of Glen Innes, and drove for about three hours. The drive started on sealed roads, but not too long into our journey they became dirt roads of ever-decreasing driveability, and eventually became undriveable. We had reached the very edge of the plateau where the streams turned into waterfalls cascading to the coast – Eastern Falls country. We parked, gathered all our gear into backpacks and began our hike down into the valley. We followed the faint hint of a trail at first, but the bush had largely reclaimed any tracks and we lost any sign of a trail after about 30 minutes. We slid down the forested slope in our gumboots, winding our way off the plateau and into the valley below, a drop of some 600 metres in elevation. Once in the valley, we walked downstream for several more hours until just before nightfall. We ate our squashed and slightly soggy sandwiches on the edge of the stream we intended to survey. I'm always most hopeful and excited when I'm at a new site, waiting for dark to fall. In these moments, at any second, I might hear the call of the peppered tree frog!

After dark, we surveyed along the stream for almost two hours. We walked together, shining our head-torches on rocks and in the surrounding vegetation, noting all the frog species we saw and heard. We found 10 species of frog, the most common being the eastern stony creek frog. This species was incredibly abundant along most rocky streams in the area. Male frogs of the species are often really obvious, turning an intense, almost fluorescent, yellow in the breeding season. Females, and many males, weren't so obvious though, being a dull brown that blended in with the rocks. They were so common that we really had to check every rock before stepping onto it. However, despite our hope and best efforts, we didn't see or hear the elusive peppered tree frog.

When the survey finished, we were still a long way from a bed. Getting down to the stream had been a physical challenge but getting back up to the car was even more so. Tim and I slowly trudged up the steep hill, checking our GPS often, to ensure that we were getting closer to the car. We had no trail to follow, and only the beam of our head-torches to light our way. It felt like we'd never get there, as we hauled ourselves up and up in our heavy gumboots, pausing to gather our breath then beginning the trudge up again. We eventually made it to the car, calves burning. At 3:00 am we arrived back at our accommodation.

The next morning we were feeling the effects of the hike. I woke up and hobbled into the kitchen to make a strong cup of tea. I knelt down to retrieve the tea bags from a container and, to my surprise, my knees gave out! I was now sitting on the floor. I concluded that we – or certainly my muscles – needed a break from endurance challenges.

Just 100 metres from where I sat on the floor was a small rocky stream, running through the rural property we were staying on. I decided that we would have an easy night and survey that stream, instead of hiking down into a steep valley again. I felt a pang of guilt, as if I was passing up an opportunity to rediscover the peppered tree frog in favour of a 'night off'. I was sure we'd only find the most common of frogs, but it would still count as a frog survey. An early night and a bit of a rest for our bodies was important so that we could continue our survey safely, but also so that we could do the surveys properly. Overtired bodies and minds tend to operate a bit slowly and miss things.

So, that night, without the need for a three-hour drive on a dirt track and with no 3 kilometre hike down into a valley, we walked less than 100 metres to the stream across flat, freshly grazed pastures. It was our 57th night surveying for the peppered tree frog. I was convinced it was also going to be the night that we would have the least chance of finding anything unusual.

As predicted, we saw and heard the kinds of frogs that you'd expect to find in a paddock near Glen Innes. Spotted marsh frogs called from flooded grassy areas, and whistling tree frogs and common eastern froglets, both almost impossible to see, were calling along the edges of the stream. All great frogs of course, but nothing out of the usual.

Rounding a bend in the stream, Tim and I simultaneously shone our head torches on a large smooth granite rock in the middle of the stream. A handful of medium-sized frogs sat on the rock. I was almost certain they'd be the very common eastern stony creek frog as they were really the only frog species in the area that sat like that, and they were so very common. However, something about them wasn't quite right.

We looked closer, craning our heads and tilting our head-torches. These frogs were slightly more rounded in body shape than expected for a stony creek frog, with blunter snouts and a less obvious stripe along the side of the face. And none were the least bit yellow. Tim and I didn't say anything for a while. We inspected the frogs intensely. Thankfully they didn't hop away.

'Are they... Booroolong frogs?' I finally uttered.

I'd never seen a Booroolong frog in the wild before. They were once common across the Northern Tablelands, as well as the Central and Southern Tablelands. However, around the same time as the peppered tree frog disappeared from the Northern Tablelands, the Booroolong frog did too. The species persisted in a small area near Tamworth on the slopes leading up to the Northern Tablelands and in pockets of streams further south, but it was now endangered and had not been recorded from the Northern Tablelands since the 1970s. It was the last frog species I expected to find. We certainly hadn't been looking for it.

We continued to stare at the frogs, scrutinising them. There was no other possibility. We were looking at the first Booroolong frogs known from the Northern Tablelands for *over 40 years*. For some reason, this species had managed to survive in this small rocky stream only 100 metres from our accommodation. Not in the most far-flung and remote parts of the forest, but in the last place we ever expected to find something this important.

For the last year or so, we'd spent well over 100 hours surveying more than 30 different streams scattered across the furthest reaches of the Tablelands. Our focus was on searching the most remote and inaccessible areas, which had necessitated the long hikes, big climbs, very late nights and jelly-like muscles. We never thought that changing our plans and looking a little closer to home would yield such an important result.

***

Discoveries can happen when we least expect them, in unexpected places and in the last places we look. If work doesn't go to plan or your original mission fails to meet its original objectives, that doesn't mean that you won't achieve something important. For around 20 years, since I first met and fell in love with the shimmering skin, delicate toes and big eyes of rainforest stream frogs, I've been working on better understanding and conserving frogs around the world. Along the way, my colleagues and I have made some exciting discoveries – from coming face-to-face with a Booroolong frog on the Northern Tablelands, to discovering species completely unknown to science in the forests of Vietnam, including a pink and yellow frog with spikes all over its back, and a tiny green frog that sings like a bird. Much of this work has required field work, often in extremely challenging conditions. Over time, I've learnt that when it comes to field work, almost nothing ever goes entirely to plan! Having a bit of flexibility and adjusting plans to, for example, avoid floods, take advantage of an opportunity or just give your poor knees a rest may allow you to uncover something important that you otherwise might have missed.

Acknowledgements

Huge thanks to the New England Tablelands community for their support of my search for the peppered tree frog, all the landholders for allowing us to survey their properties, David Coote and David Hunter from the New South Wales Office of Environment and Heritage, and all at the Northern Tablelands (Glen Innes) NSW National Parks and Wildlife Service office. This project was assisted by the New South Wales Government through its Environmental Trust.

18

Encounters with mountain gorillas

Wayne Boardman

Patches of blue sky began to break through the grey and white cumulus clouds as we ambled carefully up the porters' path towards Dian Fossey's Karisoke research camp, on the southern slopes of Mt Visoke, Rwanda. The path was well worn and, apart from some steeper gradients, was mostly a gentle ascent made difficult only by the 2,700 metre altitude. We pushed on through the lush forest to our rendezvous site at the rest meadow where we met Fundi, the main mountain gorilla tracker for Group 5, also known as Shinda's group.

It was September 2000, six years after the devastating Rwandan conflict. I was there to assess the health of one member of Shinda's group, Amahoro, a small 14-year-old blackback (young) mountain gorilla who had been ill for some time. My workplace was the Virunga Mountains in Rwanda, where the Dian Fossey Gorilla Fund (DFGF) and the International Gorilla Conservation Programme had requested my assistance.

Fundi, a gentle bespectacled ranger, was employed by the DFGF to track, observe and habituate Shinda's group to the comings and goings of people who wanted to help and study them. Like most Rwandans, he had lost colleagues and family in the recent conflicts.

Liz Williamson, the long-time Director of DFGF and a dedicated gorilla conservationist, translated Fundi's observations and answered my questions. The large smile on Fundi's face indicated that he was delighted that I had come over from Uganda to help his gorillas. Two had been lost to poaching over the last 12 months and he felt a certain responsibility, something I could easily appreciate.

Amahoro had developed a raspy cough and signs of a cold over a month ago. Others in the group had been sick, but soon recovered; Amahoro did not. Over the last few days he had lost his appetite and become lethargic, which meant it was difficult for him to keep up with his family as they moved to fresh feeding grounds. Understandably, Fundi was extremely concerned for the young gorilla.

We decided to first gather more information by observation, mull over what we had found and intervene if we felt Amahoro needed our help. This group of gorillas, one of Dian Fossey's main research groups, was habituated to trackers and researchers but they rarely saw new faces.

Slowly we walked up the slopes and through the forest into the mountain habitat of *Galium* scrub (the gorillas' favourite food) and *Hagenia abyssinica* trees, the gnarled old men of the volcanic mountains. The views to the south over Karisimbe became more expansive and beautiful. Our hearts were pounding heavily both because of the thin air and the adrenaline surge of anticipation.

Suddenly, the air was heavy with the characteristic pungent musky odour of gorilla; they were very close. We moved stealthily, muttering soft guttural sounds – *naoooooom*, a cross between a purring lion and an old man clearing his throat – to communicate to the gorillas hidden in the dense undergrowth that we were around and posed no threat.

Apparently intrigued to see their daily human companions arrive, two gorillas appeared about 15 metres away in a small clearing: a healthy vibrant silverback and a younger blackback. Within a few seconds we saw more of the group, with several juveniles playing rough and tumble, unaware of our presence. Most of the group were sitting quietly, contentedly pulling on branches and chewing the vegetation.

Liz pointed to a large, barely moving black mass lying on a ridge; Amahoro was alive but showed no interest in eating or in the visitors. Plainly, he was not behaving like the others. He was uninterested in activities around him and just wanted to rest. The dim light of the forest prevented me from observing him clearly, so we needed to approach more closely. While we were watching, Amahoro pulled himself up laboriously on to all fours and ambled off lethargically through the undergrowth. He appeared weak and listless, and his abdomen was noticeably hollow in comparison to those of the other gorillas.

As the gorillas moved on, one or two of them sauntered closer to check me out. They did not recognise this new figure and my imposing height was perhaps a little unnerving to them. More gorillas emerged from the undergrowth, peered at me and then moved closer to investigate.

Shinda, the alpha male, kept his distance, but I knew he was aware of our presence as he occasionally snatched furtive looks in my direction. Born in 1977, Shinda was once described by Dian Fossey in her book *Gorillas in the Mist* as a 'wizened, squeaking tadpole-like figure'. He lost his mother when he was only 3½ years old but had managed to assume the leadership of Group 5. His impressive and striking presence now bore no resemblance to anything remotely amphibian.

Two other young silverbacks, half-brothers Gwisa and Uegende, about the same age as Amahoro but physically more imposing, positioned themselves between me and Amahoro like conspicuous strong-armed minders. They were grooming his gaunt and weakened frame caringly, as if their attention might make him feel better. I focused on Amahoro, using my binoculars to observe his condition and decide on treatment. He had lost a lot of weight and his coat appeared duller and more unkempt than those of his family members. Frustratingly, I still I could not see his breathing rate or his face and body, which might have given me some evidence as to the cause of the problem and helped us to formulate a treatment plan.

As I was attempting to focus on one of the other gorillas to obtain some clinical data from healthy individuals for comparison, a young adult female coolly passed by me. She sat down about 5 metres away in a small clearing and keenly inspected me, this new visitor to her mountain home. She was followed by an inquisitive infant who sat even closer and stared at me. Just behind them was an adult female, probably the mother of the infant, sitting next to Shinda, who was relaxing about 20 metres away from me.

They were clearly taking a break from eating, having been actively foraging for the past four hours. Even though the young female was close, the light was not bright enough for me to easily observe her breathing rate. I focused again on her and could just see the almost imperceptible breathing movement of a healthy gorilla. As I observed her, I did not notice the infant stealthily move towards my bag and the video camera at my feet.

My colleagues had disappeared to observe others in the group and there I was, alone amid a relaxed and accepting group of mountain gorillas – or so I thought. It was an extraordinarily powerful and exciting moment, one that induced such an intense state of mind that I had a clear and undeniable feeling of 'being there' in the moment.

I heard a sound that I was unfamiliar with, a clear *hoo-hoo-hoo*. I glanced across at Shinda, who was on all fours looking away from me down a steep gully. Out of the corner of my eye, I saw him stand tall and grab a clump of vegetation. In a flash, he thumped his chest rapidly to produce the hollow reverberating *pok-pok-pok* echo that I knew to be the start of a charge. Almost immobile with fear, I realised that my presence implied a possible threat to his group. As swiftly as I could, I lowered myself down to look as small and unthreatening as possible and, on haunches, shuffled quickly behind a tiny bush. I knelt and cowered as he charged headlong towards me. I kept my eyes lowered in the desperate hope that Shinda would recognise my deference, but every muscle automatically tensed in anticipation of the crushing impact. I tried to utter an appeasement grunt, but fear had drained all moisture from my mouth. The hairs on the back of my neck stood up. Time seemed to stand still as he thundered toward me. I closed my eyes.

Nothing ... but breathing sounds.

After a few seconds I risked a furtive glance and there he was, side on to me, his steely, dark brown eyes focused on my smaller cowering frame. His mouth was partly open, displaying his blackened teeth and tongue – a clear threat gesture. Our eyes made contact briefly; his huge head, twice the size of mine, was less than a metre from me. The smell of irate silverback gorilla was strong and heady, and I could hear my heart pounding. He stood his ground, his looming black frame commanding my complete respect. My knees began to seize and sweat ran into my eyes, but I didn't dare to move. Without warning, he broke eye contact, turned away and sauntered off with a confident swaggering strut, his group duly protected. As relief swept over me, I realised how fortunate I was. I had witnessed the primal power of a remarkable creature completely at home in his mountain environment, and not died or been seriously injured! Encounters with wildlife cannot get much closer than this.

The investigation continued for another two hours. Many of the silverbacks and blackbacks were now more wary of me, regularly approaching with confidence while I made myself as inconspicuous as possible until they moved on. This behaviour continued; I was the main object of their interest for the rest of the day. Meanwhile, Amahoro, flanked by his brothers, seemed to lag further behind the main group. His minders continued to preen him and coax him along.

Amahoro remained out of eyesight most of the time and our opportunities to check him closely were diminishing. We decided that Fundi and I would approach the three boys and attempt to get a better view. We followed them up a steep gully. Fundi stepped confidently at this altitude as I scrambled madly along. We moved close to the group of three who had stopped to allow Amahoro to rest. We heard him cough several times; his breathing was noticeably more exaggerated and more rapid than that of the others nearby. Slowly, he gathered the strength to join Gwiza and Uegende, who were patiently waiting for him. As we moved back, one suddenly turned towards us and gently seized the inner part of Fundi's leg. It was not aggressive, but it definitely was a firm statement that our presence was not required. With the message communicated, they moved down the gully to catch up with the main group and were gone from us for the rest of the day.

Back at the base house, in the small north country town of Ruhengeri, we analysed our findings. Without doubt, Amahoro was extremely sick and would need urgent treatment if he were to survive. I believed he was suffering from a respiratory tract disease, likely a bacterial pneumonia, a possible sequela of the previous bouts of coughing and flu, but I could not be sure without a full examination. Our deliberations focused on how we should intervene. Throughout the evening, Liz and I discussed various possibilities and decided that if he was worse the next day, we should anaesthetise him. If he appeared slightly better or the same, we might merely treat him blind with long-acting broad-spectrum antibiotics – my least preferred option.

Early the next day we gathered veterinary equipment from the Mountain Gorilla Veterinary Project house and travelled to the village near the porters' track. We needed permission from the national park authorities to visit the mountain again to observe and possibly treat Amahoro. We also needed many porters to carry the equipment up the mountain. Several were selected and a tracker named Mathias was assigned to act as my minder. Mathias was a big burly handsome man with a rugged athletic physique, kindly caring face and deep basso profundo voice. He reminded me of the great actor and singer Paul Robeson, a hero of mine. I instantly took to him and felt confident in his protection.

We soon found the gorilla group above the rest meadow not far from where they had been the day before. We could sense them nearby, but not until we heard a commotion and much frantic movement in the undergrowth did we realise they were on their way to check us out.

Most of the gear had been left with porters further down the slope so as not to concern the gorilla group. Only Liz, Fundi and I ventured further to see how Amahoro was this morning. We spotted him on a ledge with his ever-faithful companion Gwisa. He appeared to eat some *Galium* – the first time he had been observed eating for over seven days. However, he was not eager to eat and still looked sickly and thin with a dull coat. He and Gwisa were parted from the group by about 150 metres but Shinda, who by now had decided I was just another harmless researcher, kept a watchful eye.

We discussed options and made the plan to treat Amahoro by darting him with antibiotics, with a view to him being checked by Fundi each day to see if he improved. If he appeared worse over the next day or two, then we would immediately return and anaesthetise him to conduct a full examination. Darting is very upsetting to gorillas and potentially dangerous to the shooters. Gorillas should not see the darting equipment or procedure. If they did, the silverbacks would be alarmed and feel the need to defend the group. Stealth was needed.

Two 10 millilitre darts were prepared, one containing long-acting penicillin and the other containing a drug called Baytril™. In combination, these provide a broad spectrum of activity against a variety of bacteria. Mathias, Fundi, Liz and I left the gathered group of porters and hiked back up the hill to locate Amahoro again. The group had moved on, but the tell-tale signs of flattened undergrowth and fresh dung provided a clear trail. Thankfully, at this height little high vegetation was present to impede our views of the gorillas. Occasionally I stole a glance at the stunning view of the other volcanoes that straddle the Rwandan border, Karisimbe and Mikeno, their forest-clad slopes home to most of the world's mountain gorillas.

Ntambara, one of the silverbacks, was sitting nearby and casually eating some plants, apparently unconcerned by our presence. Amahoro was also nearby but in a poor position to dart because very little of the preferred darting site – his backside – was exposed. I removed the pistol from my large bum-bag and fixed it to the pipe, which had been concealed down my trouser leg. Unexpectedly, Amahoro got up and ambled very slowly in front of us, stopping no more than 7 metres away. Immediately in front of me was the silverback's large rounded backside, fully exposed against a background of Rwanda's rolling hills. Mathias, who had been by my side the whole time, urged me to dart him. I managed to still myself before pressing the trigger on the dart pistol. The dart hit the hairy black target and discharged completely as he rushed down the slope to hide in the undergrowth. Success with the first dart! One more to go ...

Amahoro disappeared for a few minutes. Liz discovered him lying on his back in a small clearing, and intimated his location to me by pointing. I crawled on my hands and knees along a small narrow corridor of foliage until I could see his right inner thigh exposed. I drew the pistol and pressed the trigger. The second dart hit his leg and discharged completely. Amahoro leapt to his feet and moved on. He had not seen me and neither had any of the other gorillas.

None of the group seemed to be disturbed, and they continued to forage. Only Amahoro, now seeking solace from his ageing mother Pandora, was disturbed. He could not understand why annoying invisible insects had been attacking only him.

As a wildlife vet, it just did not get better than that! The darting had been a great success. We left the group and relaxed in a small clearing. Mathias and Fundi were ecstatic that we had treated Amahoro successfully, as was I. Ever watchful, Fundi was allowed on the slopes each day. He happily reported that Amahoro's appetite and activity were increasing, perhaps because of the treatment. Amahoro had at least survived another day to contribute to the next generation. When only 650* mountain gorillas are left in the wild, every one of them plays an essential part in supporting the survival of the species.

I was incredibly privileged to join the world of mountain gorillas for a moment in time. What I learnt from the whole experience was manyfold.

---

* Now 1090 mountain gorillas are left in the wild thanks to conservation warriors like Fundi and Mathias.

For example, zoo-based veterinary skills are readily transferrable to dealing with free-ranging wildlife populations, sometimes luck will put you in the right place at the right time – and knowledge of animal threat behaviours can be life-saving!! Perhaps more than anything, I learnt that we cannot save the world's endangered species without the real conservation warriors like Fundi and Mathias, passionate protectors of gorillas and their home, who tackle the difficult issues on the ground. They are the real unsung heroes!

19

Lazy lions and hungry hyaenas

Robert Heinsohn

The lioness sat up suddenly, her head rising out of the long grass. She stared at me inquisitively with large green eyes, neither alarmed nor frightened. I froze as a surge of adrenaline flooded my body. Fight, flight or freeze. Never had I been so alert. I was frozen in place for what seemed like an eternity, but conscious thought eventually returned. I was by myself, on foot, in the long grass of the Serengeti Plains. Barely 20 metres from me was a very large predator. My peripheral vision and a lightning-fast mental calculation told me that the nearest tree was at least 200 metres away, so it was pointless to run for cover. Then, to my horror, a second head appeared off to one side and slightly behind me, then another on the other side. The three lionesses had formed an equilateral triangle with me at its centre. My actions took on a life of their own. Almost without thinking I began to back away, oh so slowly. One foot and then the other, never taking my eyes off the first staring lioness, I angled my body between the other two lionesses. My instincts were guiding me slightly closer to one of them because she was more concealed by a tuft of long grass than the others. Maybe if I couldn't see her so well, she would be less inclined to worry about me? Eventually (I can't say if it was three minutes or three hours) I crossed the edge of the triangle, turned and subtly quickened my pace to a slow walk. I wanted to run but by now I was in conscious control of my actions. I forced myself to walk towards the trees, keeping watch over my shoulder. When I had put about 60 metres between us, the first lioness rolled over and flopped down in the grass. Her buddies did the same. It seems they were not in hunting mode that day and I was not worth eating, this time ...

I had arrived in the Serengeti about six months earlier and would end up 'postdoccing' in East Africa for almost four years. To say I was young, naïve, immature or wet behind the ears would all be understatements.

Although fairly outdoorsy, a bushwalker and generally capable, I had never lived anywhere so remote. I had always had a strong drive to live somewhere truly off the grid after my PhD, and wrote several letters to professors who ran field projects on beautiful animals in spectacular places. I argued that my background on highly social bird species made me an ideal candidate to do research on group-living mammals. Two scientists bought this argument. One worked on primates in the Amazon and the other on lions in East Africa. For complicated reasons, the planets aligned for the latter and in 1990 I made my way to Tanzania.

The Serengeti Research Institute is in the middle of the Serengeti National Park and surrounding game reserves, a combined area the size of Belgium, located in northern Tanzania. It is vast and beautiful and, and back then at least, totally unspoiled. It was also very isolated in those days. No phones were available, and email hadn't been invented. The nearest town was Arusha, an intensely bumpy and tedious seven-hour drive away, and with limited goods or services once you reached it. Nairobi, the big smoke up in Kenya where you could get essentials, was a 16 hour drive. The name of the game in the Serengeti was self-sufficiency. Bring in all your food, spare parts and sundries for at least two months at a time, and survive. You could drive to town more often but doing so inevitably led to mechanical problems as something on the car was bound to break *en route*. The first week after a visit to town was pleasant enough as the race was on to eat all the fresh food before it spoiled, but after that life got grimmer and grimmer with nothing but rice and beans until the next trip. The steepest learning curve was vehicle maintenance. I wasn't one of those kids whose parents let them play under the bonnet – more the opposite, in fact. However, I was soon a dab hand at changing the leaf springs of my little Suzuki 4WD, and not too bad at cracking split rims to mend a puncture.

So, what was I doing surrounded by lionesses in the middle of nowhere? I had been recruited to work as a postdoc for the Lion Project, which had been running for about 30 years by then. For the previous 12 years or so the head honcho was Craig Packer, a professor from the University of Minnesota, who brought me in for a project to work out the whys and wherefores of cooperative territoriality between pride members. Lions must live in groups (prides) to hold a territory and raise their young safely. They work very closely together to achieve these aims. How they manage to cooperate and coordinate so well with one another was a mystery with broad ramifications for understanding group living across all social species, including humans.

After leaving me to flounder in the Tanzanian bureaucracy for the first three months (including many days of sitting outside the government minister's office in Dar es Salaam), Craig eventually turned up in the Serengeti to get me started on the project. As we drove around the 2,000 square kilometre study area, tracking the lions via their radio-collars, Craig shared some of the insights of his long experience. These pearls of wisdom included how to navigate the pride territories, identify individual lions by their whisker spots, dart lions to change the radio-collar – and lots of advice about staying alive. One of the phrases that stuck in my memory was 'Never leave the car!' In other words, if you break down, wait for someone to come and find you. It was simply too dangerous to try and walk back across country given all the animals that could charge (buffalo and elephants), bite (hippos and puff adders) or eat you (lions, hyaenas and crocodiles). The Lion Project usually had two researchers, each operating independently but able to look out for each other. The idea was to tell each other where you would be going that day (or for how many days), and if you hadn't turned up by the scheduled return time your the other person would come looking for you. Unfortunately, the radios that would have helped the process had burnt out long ago, but at least a system was in place. The only trouble I had with telling my colleague where I would be was that I didn't have a colleague at that stage.

One day, I was bumping along in my good old 'Suzi' heading for home after a morning looking for the Masai Kopje Pride with which I was particularly keen to do a playback experiment. These experiments entail placing a speaker near the pride and playing the roars of strangers, which tricks the pride into a territorial response as they think they are being invaded. Playbacks are lots of fun as the lions (often seen lazing around during the day) become very alert and assertive in their behaviours, thinking they are heading into battle. The process produces great video footage with many data about individual roles in the pride and their keenness to act. I had found the pride and they were in a good position for a playback, but I still had a few hours to kill before sundown, after which I could start the process. So I went looking for more lions. The more often that each pride was spotted, the better the long-term demographic data (i.e. information on which lions were still alive, who had cubs, who was cavorting with who) collected for the whole project. I drank a cup of lukewarm tea from my broken old thermos, marked the location of the pride on the map (no GPS back then, as the US military was still scrambling the signals) and cruised off across the plain.

Long grass plains are easy to navigate except for one thing – big holes made by warthogs, aardvarks and hyaenas! An almighty crash followed by an even bigger bump, with everything in the car tossed like a salad, told me that one of my front wheels had gone down a hole. Not to worry, this situation happened all the time. It was just a matter of using the high-lift jack to lift the wheel out of the hole and back onto *terra firma*. Except on this occasion, I had left my jack back at base. I pondered what to do given that no one knew where I was, and base (the Lion House) was only about 5 kilometres away. It was before 3:00 pm, the landscape between me and base was all open grass plain, surely I could make it? Then I met the lionesses ...

Nothing reminds you of our evolutionary history more than coming face-to-face with a large predator. My study area was located only 40 kilometres from Olduvai Gorge, the famous site of Leakey's discovery of early *Australopithecus* specimens. I liked thinking about early hominids like Lucy wandering across the grassy plains under the shadow of an erupting Mount Kilimanjaro. After my encounter with the lions, I could literally taste the fear of my hominid ancestors as they watched warily for big cats and hyaenas. I could also understand their intense need to band together in cooperative groups for protection, hunting and raising children. I loved this sense of walking (although usually I was driving) in the footsteps of my ancestors. It was a sobering reminder that we're just another animal in the food chain, and despite our clever ways of keeping nature at bay, we can still end up as prey if we let our guard down.

One fact I am sure of is that my ancestors, if given a second chance, would have learnt predator avoidance much more quickly than I did. The need to dodge predators is one of the presumed reasons for evolving ever larger brains. My brain is larger than that of *Australopithecus*, which is why

I am at a loss to explain why I left the car for a second time, just a few weeks after the lioness incident. The car broke down this time because of a fuel blockage. It was harder to work out exactly where I was, but I calculated that I was probably about 10 kilometres from Lion House. It was 4:00 pm, sundown was at 6:00 pm or so. I had two hours – surely that would be enough. No one would be coming to find me, and I was hot, dusty and tired. I certainly wasn't in the mood to spend a night in the Suzi.

So I started walking back to base. Again. Despite having been told not to. This time I had a vehicle track to follow, so I was less likely to walk right into sleeping carnivores. I made good progress and felt confident. At around 5:00 pm I looked over my shoulder and saw that I had a travelling companion, a spotted hyaena roughly 100 metres behind me on the track. Oh well, I thought, not too much to worry about if it keeps its distance. I kept going, but the next time I looked back two hyaenas were following. And they were a bit closer, ambling along in that unmistakable hyaena gait about 80 metres behind me. 'Hmm', I thought, 'I should be okay as there're only about 3 kilometres to go and it's still broad daylight.' A short time later, three hyaenas were on my tail and they were closer, maybe 50 metres behind me. I decided to do something before the situation became trickier. I ran at them, screaming and shouting. They scattered and I resumed my very (!) brisk walking pace. The sun was low on the horizon, and although not sure exactly of my location I knew by the lie of the land that I was approaching the plain of Lion House. 'Still maybe 2 kilometres to go?' I pondered. Five hyaenas were now following me (they had quickly regrouped after my attempt to discourage them), the sun had dipped below the horizon and I was beginning to feel seriously scared. I could see a few small trees in the distance that I could possibly scale if desperate, but I really wanted to reach the four safe walls of Lion House. I calculated that camp was still about 15 minutes away. I ran at the hyaenas once more, and they scattered – but not very convincingly. They closed to within 30 metres of me. In desperation, I ran at them again. Soon after, I came around a bend and crossed a small creek. As I climbed up the bank I knew where I was. I could see the old airstrip and just beyond that was Lion House, situated on the edge of the scattered buildings of the Institute. I had 500 metres to go. I yelled and screamed, and semi-scattered my persistent hyaena shadows one last time. I started jogging, counting down 400 metres, 300 metres ... At 200 metres I let loose and sprinted the last stretch along the dusty track. No one was timing me, so I'll never know if I met Olympic standards that day, but I do know that I moved faster than I ever have before. I also don't know if the hyaenas actually chased me, snapping at my heels, but I imagined I could feel their hot breath behind me and could virtually feel their paws thudding on the ground in pursuit of a meal. I do know that every fibre of my being carried me to the front door of the house. Not everyone gets a second chance in those kinds of situations, and although I had other close calls during my stint in East Africa, one thing is for sure – I never left the vehicle again.

Acknowledgements

Thanks to Craig Packer for giving me the incredible opportunity to study lions in the Serengeti. It was not always easy or comfortable but I can honestly say they were the most amazing and formative years of my career.

20
Don't be a lemming

Jo Isaac

It was the year 2000. New Year's Eve 1999 had been and gone, and the Millennium Bug was one of the biggest non-events of the century. I was relieved, as multiple copies of my Honours thesis (or dissertation) were saved on the most advanced floppy disks of the day, which I would have been devastated to lose to Y2K problems!

The Honours research in question had investigated predator–prey cycles, specifically those arising from the interactions between field voles and weasels, on the Scotland–England border. Voles are rodents from the subfamily Arvicolinae, which is the same animal group that lemmings, those fluffy originators of the mass-suicide myth, belong to. Voles are herbivores who mostly munch on grass, and are important prey for a variety of predators. Indeed, I once heard field voles described as the McDonald's of the forest – a tasty takeaway morsel for any undiscerning carnivore! The UK has three species of vole: field, bank and water voles.

Field (or short-tailed) voles are tiny things, averaging 10 centimetres in length and weighing in at 30–40 grams. They are as common as the proverbial in their favoured habitat, and conveniently leave signs of their presence in the form of runways and grass clippings. Chunkier at up to 300 grams, water voles are chubbier, cheekier and unfortunately rather less common than their short-tailed cousins. Classed as endangered in the UK, water voles are affected by habitat loss and fragmentation, as well as predation by the introduced American mink.

Following my graduation in July 2000, my Honours supervisor employed me as a field coordinator, which involved conducting field work on field and water voles. My water vole field work primarily took place in northern Scotland, in the Cairngorms National Park and along the burns (a colloquial name for a small stream, brook or creek) and waterways of the wild West Coast. The goal that summer was to resurvey known colonies to determine if they still existed. Water voles in Scotland tend to exist in metapopulations. For possibly the most expressive description of metapopulation dynamics ever written, I will borrow a couple of sentences from an ecologist infinitely more poetic than I, and quote from a 2005 water vole report by Rob Raynor:

> Metapopulations are analogous to a constellation of twinkling stars blinking on and off. Some stars may dull (local extinction) while others appear (local colonisation). Thus, while the pattern of reflected light (animal distribution) changes over time, the overall stability of the spectacle (regional population) is maintained.

Essentially, water voles live in colonies dotted along a burn or other watercourse. Breeding females defend territories of around 100 metres in length, while males have home ranges of up to 300 metres. Surveying the area involved finding and then following the watercourse, and looking for signs of voles, including burrows, latrine sites and/or piles of grass clippings left from lunch.

When we found a new water vole sign, we'd set small aluminium box-shaped traps nearby, bait them with carrot and check them morning and evening. The traps had an opening at one end for the voles to enter, enticed by the tasty carrot treat. They would have to cross a plate with a treadle mechanism underneath, which would cause the door to slam shut and trap the creature inside. Overall, it was pleasant and rewarding work.

The Scottish weather can be fickle, and a raincoat was essential. Even in summer, the temperamental Highland weather can take you by surprise. In the weeks preceding one of our more eventful field trips, everything from snow on the high peaks (in summer, mind you!) to sweaty humid heat had occurred. Also, the mental anguish of being bitten by swarms of the infamous Scottish midges can't be overestimated. However, by and large, the walking was easy and through a combination of old-school Ordnance Survey (Great Britain's national mapping agency) maps and a trusty old yellow Garmin eTrex®, we found our field sites.

Our study area at the time was in the Cairngorms National Park. We were relatively isolated, with no vehicle, an unreliable satellite phone and a solid three-hour walk to the nearest National Park office. A local wildlife consultant, Pete, was also in the general area conducting his own surveys and knew where we were camping and working.

Our days began much like every other, with my field work partner Lisa and I waking in our respective one-person tents. We'd pitched them next to a small ruined stone hut – known in Scotland as a bothy – which we used for shelter and cooking purposes. Morning ablutions were a rudimentary affair, usually involving a short walk with a shovel and a one-way conversation with the local mountain hare, who had an intense and somewhat unsettling stare. After breakfast, we set off on our morning rounds to check the water vole traps. The primary watercourse in the area was the mighty River Dee, but our traps were set on smaller nearby burns. To access our trap route, we had to cross one of these small burns, but that summer the water was so low that you could walk through it at the vehicle crossing point and barely moisten your gaiters.

The vista and terrain in the area were very typical for the Cairngorms – low vegetation of grasses and heather, very few trees and mostly rolling hills, with a few higher craggy peaks. Burns trickling with peaty, tea-coloured water twisted around the landscape, and a few larger murky lochs were dotted about. Given the proclivity of the voles to be found around water, traps are usually set quite close to the water line on the bank of a small burn, near vole signs and traces.

Our morning trap check usually took a couple of hours, and this day was no different. We walked a loop, processing any animals in the traps and releasing them at the point of capture. Water voles can be active day and night, so we reset the traps to be checked again in the late afternoon. Processing a vole with two people was a relatively quick affair – a hair sample, body mass and other measurements were taken, and a uniquely numbered ear tag was attached if the vole did not already have one. Then we'd release them to go about their business.

Back at camp by the early afternoon, we settled in for afternoon tea and began sorting out our data, maps and field gear. When it began to drizzle, we relocated inside the bothy with our steaming tea and delicious carby snacks. As we chatted and worked for a couple of hours, the drizzle seemed to increase in intensity. In truth, we were probably way past a 'drizzle' by then – it had become a consistent downpour.

It was a couple of hours before our next trap check and I was mostly focused on taking a short nap in the tent before we had to set out again, but I eventually wandered outside to assess the weather. As I glanced up the valley, I noticed something odd I'd never seen before – a waterfall appeared to be falling from one of the higher rocky peaks. I called Lisa to be sure that I wasn't hallucinating, and she quickly confirmed my observation. A new, rather large and ominous-looking waterfall was gushing water into the valley below. The valley where we were standing and where our traps were set. Naive and still not overly alarmed, I suggested we check the nearest burn to see if the water was rising.

Our usually toe-deep river crossing was solidly above waist height and flowing at a fair pace. Murky peaty water crashed and frothed downstream, taking with it large debris such as branches and sticks. The gravity of the situation finally dawned on me. Any water voles that might have been caught in our dozens of traps set very close to the rapidly rising burns over the last couple of hours were at risk of drowning.

***

Allow me a slight digression to explain, briefly, how I ended up as a zoologist wading through the Scottish floodwaters ... Like many (but not all) zoologists, I've loved animals since I was a wee bairn. However, I did dreadfully at school, passing few exams and leaving with zero prospect of getting into university. Fast forward to my mid 20s, and after a few years of backpacking and travelling the urge to do something real and tangible for wildlife and conservation had become almost overwhelming. I wrote to Greenpeace. I wanted to save the whales, save the pandas, hug the trees. They told me to get a degree in zoology. So I did, about five years later. My passion for animals, wildlife conservation and halting extinctions is what drove – and still drives – my career choices. Thus, I could not entertain the idea of being responsible for the loss of a single water vole. Lisa was of the same opinion.

***

We grabbed our gear, put on raincoats and headed off. Our first obstacle was the previously ankle-tickling burn on the way out of camp. It was now a raging torrent of angry brown froth, littered with bobbing sticks, branches and the odd floating deer carcass. Until now, our concern had been for the animals' welfare alone, but now we had to enter the water. At this point, it became clear that the raincoats were somewhat pointless. We navigated the bank, and I gingerly slid in first. Being several inches taller than Lisa, we reasoned that if it was too deep for me then she would certainly be swept away, and we'd have to find another place to cross.

The water was about up to my chest, which was about equal to Lisa's chin, but we held our breath, clasped hands and started wading. We both began laughing hysterically, which was more of an anxious cachinnation. However, it helped us navigate the crossing and we ungracefully hauled our wet selves out onto the opposite bank. We set off hurriedly to get around the trap circuit. We had caught some water voles in the few hours since our morning check and all individuals were fine, if a bit damp. Processing and data collection went out the window – we were focused on search and rescue. The nearest miss was one female vole whose trap was half underwater by the time we got to her. She was a bit bedraggled, but was okay and trundled off on release. After clearing and shutting all the traps in record time, we headed back to camp.

Our plan was to cross the burn in the same spot, but while we'd been clearing traps and rescuing waterlogged voles the rain had continued in a determined, very Scottish, way. The burn had practically doubled in size. Our original crossing point, sketchy in the first place, was totally out of the question. We estimated that the river was well over 2 metres deep and the debris flowing down it was a serious danger. The gravity of the situation took on a new slant – while the animal welfare calamity had been averted, early career zoologist safety was now in jeopardy.

We considered trying to walk to the local National Parks office, but there were several other water crossings along the way whose depth we would not be able to gauge. We could sit it out, but we were already soaked to the skin, cold and had no idea when we might be able to get back to camp, particularly if the rain continued. Neither of us fancied the idea of a cold wet night when our dry clothes and tents were just across the river! So, we decided to try and find a safer crossing point. We squelched up and down the burn – I had never appreciated the serpentine path it took as we tried to follow it looking for a shallower, narrower or slower section!

Using sticks as rudimentary depth gauges, multiple attempts were aborted as we got colder and more desperate!

After an indeterminate amount of time (it felt like hours but may well have been more like 45 minutes) we managed to cross the raging burn to relative safety. I don't have much memory of that crossing, aside from knowing that we made it. Relief washed over us. We weren't out of danger yet, but reaching the comparatively dry bothy produced a short euphoria. We planned to get dry, attempt to use the unreliable satellite phone and wait for help. We had enough pasta to last us at least a couple of days.

However, only a few minutes had passed when a 4WD came roaring through the churning black waters. Pete had realised that we might be stranded and came to check that we were okay. He was (rightly) concerned that even a short delay could mean the difference between getting the car back across the flooded ford – or not. Chaos ensued as tents were yanked down wildly and thrown in the back of the car. Lisa and I jumped in, still in our soaking wet clothes, and Pete expertly navigated the floodwaters, driving us all to safety.

***

It's been nearly a quarter of a century since this voleing escapade and, as is often the case with field work mishaps and adventures, it has become a tale of legend. My 16-year-old son has heard it so many times that he can probably repeat it verbatim (I still made him proofread this chapter). Over the years, other, mostly less serious, field work misadventures have happened – some involving voles and some not. This one is embedded in my psyche for several reasons that may be helpful for those starting out in, or considering, a career in ecology. First, it was my first really frightening (more so in hindsight, less so at the time) field situation. Second, it happened before more reliable safety and communications methods were widespread – looking back, we really were conducting the work in a more old-school fashion. It often strikes me how much safer we are with all of today's technology. It was also the first time I really became aware how quickly conditions can change in the field, and how dangerous that can be. That is knowledge I have taken forward and used in all my field work ever since.

Even though it was a risky situation, many of the actions we took helped to keep us safe. We stuck together – it could have been tempting to split up after the first sketchy river crossing, with one of us going for help and the other shutting the traps. That decision could have been disastrous, as we had no way of contacting each other. Also, even though our communications were sketchy (a product of the times rather than anyone's fault), people did know our location, and that we were out there on our own with no vehicle. Obviously, Pete knew where we were, but so did National Parks staff and our employer. It's safe to say that even if the satellite phone failed us, someone would have eventually rescued us if we stayed put, which would have been our safest option once we'd got back to camp. Making sure people know where you are, having a clear line of communication for emergencies and remaining calm (sometimes easier said than done) become critical in these types of situations.

I will close with one final, tangentially related and scientifically questionable tale. As I mentioned, voles are closely related to lemmings, about whom there is a persistent myth of mass suicide by jumping off a cliff. Of course, lemmings don't commit suicide, either individually or *en masse*. Several scientific papers have tried to rationalise how this myth originated. Like field voles, lemming populations display cyclic population dynamics, and during a boom phase (when numbers are very high) large groups of lemmings can disperse to look for a new home. In 1996, Kock and Robinson hypothesised that, during dispersal, lemmings may try to swim across rivers and lakes. These ecologists recorded high mortality rates of lemmings as a result of drowning. Thus, it is thought that early observers found the tiny lemming bodies littering the banks of waterbodies and 'jumped' to the wrong conclusion. I feel that if a lesson is to be learnt from the misadventures of myself and Lisa, perhaps it is 'don't be a lemming'!

Acknowledgements

I'd like to thank Professor Xavier Lambin for taking me on as an Honours student and introducing me to Vole World, and subsequently employing me in the position that led to this adventure and many others. Thanks also to Lisa George for being part of the adventure, and for fact-checking my story to ensure I hadn't embellished anything over the interim 23 years!

21

Places where a kea's beak shouldn't be

Lydia McLean

Kea – those large, charismatic and highly social parrots found only on New Zealand's South Island – are known for sticking their beaks into places they shouldn't. They are the only alpine-dwelling parrot in the world and are exceedingly intelligent. Whether tearing into tents, undoing window seals or pulling the stuffing out of a tractor seat, everyone who visits kea country has a hilarious or infuriated story about their antics. Animal cognition researchers have ranked kea as one of the smartest non-primates in the world. Kea can make statistical inferences and solve tasks involving multiple types of tools. With such a remarkable brain, they are adapted to survive the harsh conditions of New Zealand's mountains, forming large social flocks and working together to seek out novelty and entertainment. Their broad generalist diet allows them to find food in unpredictable places, from forested areas at sea level to the subalpine zone among the tussock grasses and scrub. They commonly forage for berries, roots, invertebrates and carrion, and less commonly consume junk food, attack windscreen wipers and steal socks.

In a modern human-dominated world, keas' curiosity frequently gets them into trouble as they investigate vehicles, toxins and structures that would be better left alone. I have the privilege of working with kea in their wild mountain home, researching their diet and behaviour as part of my PhD while also conducting field work for the Kea Conservation Trust and New Zealand's Department of Conservation. Working with kea poses several logistical challenges, not only because they live in the untamed wilderness of New Zealand's Southern Alps but also because they will go to great lengths to sabotage attempts to study them. Let me give you a few examples.

Once bitten, twice shy

The kea's beak was clamped firmly to that fleshy area between my thumb and forefinger. I winced as he wrenched my flesh from side-to-side, macerating it and drawing blood, as I grasped wildly with my other hand for something to prize him off with. A kea beak is like a multitool, adapted for just about every purpose – including chomping on a hand. Normally, an experienced kea handler can avoid bites, but on this day I was somewhat distracted. Our team of two rangers was run off its feet, bombarded by a raucous group of juvenile kea which had shown up unexpectedly to investigate our mountain campsite. Earlier in the afternoon, we had had naively thought it safe to erect the tent fly, which was now suffering death by a thousand kea bites. We were attempting to catch female kea to attach radio-transmitters to, so that we could track them to their nests as part of a study to understand patterns of nest predation (lots of exotic species, like stoats and cats, prey on kea eggs and chicks). We were also applying coloured leg bands to each bird we caught so that individuals could be identified from a distance, allowing people to report sightings of known individuals. My teammate had his hands full with the first bird. I was charged with banding while also making futile attempts at crowd control with the remaining flock. As soon as I focused on the delicate task of banding, the other 12 or so kea would take the opportunity to cause havoc, stealing any loose items they could find and pecking holes in pieces of equipment left unattended for more than a few seconds.

Kea are caught for banding by using small snares made from fishing line attached to a coffee sack. The hessian sack is firmly tethered to the ground. By playing pre-recorded calls and putting brightly coloured objects in the middle of the snares, the kea's curiosity eventually wins, compelling it to waddle up and snag its toes while investigating the novelty lure. Kea in the remote backcountry are often more cautious than those which have experienced humans before, because they haven't learnt to associate humans with food. However, mob mentality had taken over this group, and they egged each other on to get as close to our gear as possible. While we regularly go for many days without seeing a kea in the backcountry, when it rains, it pours. We were catching kea at such a rate that we had to let several go without banding them because we couldn't keep up. Undeterred, we recaptured several of these individuals later. Knowing what kea are like, they probably saw being held and released as some sort of cheap thrill or game.

Back to my macerated hand – the banding pliers were the first thing within reach, and I wriggled them into the kea's beak in place of my bleeding skin. He proceeded to strip the rubber off the handles. To distract the others, I tossed them an empty tin can – the only item I could find that would not cause them any harm. Like seagulls to a chip, they crowded around the can, jostling for a turn to nudge it with their beaks and roll it across the ground. It was like a game of soccer for little kids, with everyone crowding around the object of interest, chasing it as it rolled around and competing for their moment of glory with the coveted can. One of the birds put his head inside it and stood up with the can on his head, stumbling around completely blind. However, before long the can came off his head and rolled off the cliff, whereupon the flock of kea resumed their task of poking multitudes of individual beak holes into my tent fly. With a bit of strapping tape hastily stuck over the bite in my hand, we continued to catch and release 12 kea that evening. A win for our study, but not for my hand or my tent.

A thieving beak

We stepped out of the helicopter and crouched over our bags, shielding our eyes from the rotor wash as the chopper lifted off and receded into the distance. We straightened up and took a few seconds to appreciate the wondrous silence of the mountains. Our site was on top of a high bluff that we'd visited before. Kea seem to enjoy this place, swooping on the thermals and base jumping off the edge. On cue, we heard a familiar cry as a kea swooped in to check out the new arrivals in his valley. Unlike almost every other bird, kea are attracted to noise and will investigate its cause. Often, we can call kea by creating a din of banging pots and pans, or by rolling rocks down the mountainside. These methods are typically just as successful as the more traditional method of playing kea calls from a speaker (call playback). Similarly, kea seem to enjoy creating noise themselves, being compelled to announce their presence brazenly with a shrill cry as they arrive, settling on a perch to overlook the novel scene before them.

It was the start of a 10-day trip to monitor blood lead levels in backcountry kea. Lead poisoning is a relatively recent finding in kea populations but has probably been occurring since Europeans first settled within their range and started constructing buildings with lead components. In a paper from the 1960s, a researcher described a kea exhibiting 'psychosis'. In hindsight, the behaviours he described are consistent with the symptoms of acute lead poisoning: loss of coordination, wing droop and emaciation. Lead is both malleable and sweet, and is used on hut roofs in the form of nails and flashings. Kea chew on the soft lead parts of these structures, unknowingly consuming a toxin that affects every part of their body. Like many birds of prey, kea also scavenge on carcasses left behind by hunters, and consume fragments of lead ammunition from the bullet tract. The geographical range of lead poisoning appears to be focused on a few hotspots, and our job was to test blood lead levels in a new area.

To sample blood, we must catch a kea and withdraw a tiny sample from the vein under the wing. A drop of blood is carefully measured into a capillary tube then put into a buffer solution for analysis. Capillary tubes are a critical part of the kit, and we cannot test for lead without them. They are kept in a cardboard tube about the size of an empty toilet paper roll – as we discovered, the kea consider this a very stealable-sized object. A moment's inattention saw the tube roll across the ground away from me. As if in slow motion, a kea hopped sideways in that goofy dance they do and scooped the tube up. He triumphantly took flight and headed out over the edge of the cliff with the tube in his beak. While making eye contact with me (I'm sure he did) he let the tube go and cocked his head sideways to watch it fall into the dense scrub below.

With no backup lead sampling gear and nine more days before we were scheduled to be picked up, we began to search for the tube. After several hours of cursing, scratches and wrestling with very sharp bushes we managed to find it. I have been much more careful with my kit ever since.

Camera interference

It had been a long day to find Isobel's nest. I'd already made a couple of attempts at finding her by scanning for the small radio-transmitter we'd attached to her some time ago. I'd traipsed up several thousand vertical metres over the past few days, chasing false leads through bush and bluffs, finding an alternative path because of incised canyons on side streams, getting tangled in thick undergrowth, getting misled by transmitter signals bouncing off the many small cliffs in the valley. Confident that I was on the right track this time, I shouldered my pack, lifted my telemetry aerial and headed up the mountainside towards the faint beep coming from her transmitter.

Kea almost always nest below the treeline, within the relative shelter of the forest. However, Isobel was a rebel. The beeps from her transmitter directed me ever upwards; I crossed out of the forest and into subalpine scrub. Higher and higher I climbed, clawing for purchase on tussock grass as I scrambled up the side of a cascading waterfall. Eventually the beep became sharp and clear, and I dropped my pack with the knowledge that I should be able to locate her nest in the next 50 or so metres. I pulled myself up on a thick stem of subalpine scrub and spotted a male kea outside a cavity in the rock. Finally, after days of searching, I had found Isobel's nest.

A quick poke around the nest confirmed that it was occupied. Although I couldn't get my head inside the cavity, I employed a highly sophisticated backup tool: my phone on video mode strapped to a stick with my headtorch. The blurry footage mostly consisted of moss and rocks and dirt, but for a split second I had a clear shot of Isobel and her three gangly chicks. That was all I needed to know. To work out if the chicks survived and managed to fledge, I installed a motion-activated trail camera outside the nest to monitor for intrusion by stoats and cats. The fate of each nest is just one data point in a study of survivorship, but these are hard-earned data points. It had taken me days to find this nest and I would need to revisit it several times over spring to collect the cards from the camera, review the footage and confirm the number of chicks left inside the nest.

Three weeks later, I was back to check on the nest. I arrived to find the camera's hatch wide open and all the control buttons chewed off. The memory card, which should have been sitting in the slot on the base of the camera, was nowhere to be seen. I had forgotten to put a hose clamp around the camera on my last visit – a very simple device that holds the door shut and prevents kea interference. Defeated, I attempted to search the surrounding area but the steep terrain made it difficult. I hung on to whatever plants would hold my weight with one hand, while sweeping the leaf litter with the other. After a while, I laughed at the futility of the search and gave up. I'd never find a tiny memory card on this mountainside, and the kea had probably taken off with it anyway. I began the trudge down the hill, berating myself for being an idiot and vowing to never forget a hose clamp again. I turned backwards to downclimb some roots – right in front of me, like the proverbial needle in the haystack, was the memory card. Elated, I scooped it up, placed it safely in my zipped pocket and began the scramble back down to the hut.

Later that night I reviewed the footage. After some beautiful pictures of fledglings learning to stretch their wings, the last photo looked straight down a kea's open beak!

Designer boots

Shy Lake is in the remotest part of New Zealand, located in the south-west corner of Fiordland National Park. It takes about 40 minutes to fly there by helicopter and would take at least two weeks on foot. We were there to apply transmitters to kea as part of a study of their survival following a poison-baiting operation intended to kill stoats and protect a rare species of kiwi in the area. There is a small research hut at Shy Lake, which is the only human structure for miles around, and we had hunkered down inside to see out a spell of terrible weather. On the afternoon before the rain arrived, we took a final opportunity to catch a pair of kea sitting in a tree nearby. After spending several hours playing calls while dancing and singing (they particularly like Madonna), banging pots together and putting out colourful lures, they still wouldn't come down from the tree. Twilight arrived and it was time for a cup of tea. We packed up our gear and went inside to put the kettle on.

About 20 minutes later, I needed to go out for a pee and headed for the communal pair of gumboots so I wouldn't get my socks muddy. The gumboots were gone. I was sure they had been at the door. After more consideration, I affirmed that yes, they had definitely been left at the door ... outside the door.

Sitting on the doorstep, almost as if to taunt me, was a wing feather. A calling card! Flapping noises from the bushes gave away the kea's hiding spot. I took my socks off and went down the hill barefoot to see what was going on. The startled kea flew away, leaving behind the gumboots, neatly cut short around the tops and turned into clogs! These kea had beaten us at our own game – they'd made us dance around like lunatics all afternoon, no doubt entertaining them in their high tree perch, then came boldly and silently to the hut door, grabbed the gumboots with their thieving beaks and snuck away. The gumboot incident occurred several years ago, and the clogs remain at the hut to this day as a reminder that kea will always get the last laugh.

Backseat driver

The timing was perfect. As we entered the winding 'No Stopping' section of the highway, the kea swiftly popped off the door of the carry cage and hopped out onto the back seat.

'He's loose, pull over', I instructed my friend. 'I can't!' With blind corners in front of and behind us, we were forced to continue driving until we could find a safe place to stop.

We had collected the sick kea after a call from a local ranger, who had been testing blood lead levels in many of the birds found by the roadside in Fiordland. This particular kea had returned an alarmingly high lead result and needed to be immediately taken to the vet for treatment to remove the toxin from his system. I was in the area, so I volunteered to have a look for him. He was easy to find, staggering drunkenly around the roadside with such impaired coordination that I was able to catch him easily by hand. I placed him gently on a towel in a cat carrier and strapped the carrier into the back seat of the car. It wasn't the ideal setup, but it was all I had for this last-minute call-out. I wasn't wearing identifying clothing that identified me as a legitimate researcher and was driving my beat-up Subaru, so I attracted some concerned looks from nearby tourists. Assuring them that the kea's capture was a legitimate operation, we headed back towards the town.

The flimsy plastic knobs on the carrier were no match for the kea, even with his reduced capacity. We'd only driven a couple of minutes when we heard the door ping off, and I turned around to see him sitting loose on the back seat as if he were my dog and we were on the way to the beach. The narrow road was shouldered by a steep bank on either side, and a series of blind corners meant we had to keep going. By the time we could stop, he had made himself comfortable, depositing a large purple poo and positioning himself side on, as if ready for any further attempts at handling him. Very reluctant to open the car doors – which might allow him to escape – I wedged myself awkwardly between the front seats. I would usually pick up a kea from the back of its head to avoid the sharp beak, but that wasn't an option in this situation. I winced as I sacrificed a finger to the waiting beak and we jiggled him back into the carrier, then reattached the door as best as we could. For the rest of the journey, he was kept under close watch in the footwell of the passenger seat to prevent any more door removals. I was very glad to pass him over to the vet.

***

Keas' intelligence and propensity to put their beaks in places they shouldn't is one of the reasons that they are so rewarding to study. When they are pulling seals off car windows or poking holes in tents, they are simply showing the behaviours that have formed through their evolutionary history and that they are adapted to do. It is up to humans to ensure that the places within beak's reach will cause them no harm. Although most kea stories end with a laugh, we must keep a step ahead of these fantastically clever animals to prevent harmful situations arising. For example, feeding by humans is one of the most common yet preventable dangers kea face, not only because human food is bad for kea but also because these interactions change their behaviour. 'Junk food' kea, as they are known, hang around carparks and are exposed to more danger than kea that live safely in the mountains. These junk food kea are used to eating weird things and are more likely to consume poison baits that are targeted at rats and possums. Kea are natural comedians, but they also have a softer side. They build strong and enduring social bonds with each other and put a lot of time into caring for their young. It is a privilege to spend time with kea in the wild, travelling to far-flung reaches of New Zealand's wilderness to observe them in the places where they belong.

Acknowledgements

Kea are only handled or banded for specific purposes, and we agonise over the necessity of understanding their behaviour that leads us to interfere with their lives to any extent. Kea may only be handled with the appropriate licences and the permission of local iwi. I am grateful to the various Rūnaka ō Ngāi Tahu for their support of kea conservation. Hopefully, one day we will again see these taonga flourish throughout their natural range.

22

Coral slime, feisty fish, shark encounters and the importance of looking up

Tracy Ainsworth

Corals have fascinated me since I first encountered them and I remain besotted with them almost 20 years later. Corals and coral reefs are exactly what you'd imagine them to be – they are full of colour, vibrancy, action, life and death and everything else in between. Once you start to look at a coral reef you cannot look away. Scientists fall in love with them and can spend their lives following the life of a single reef. Coral reefs are everything and in this tale from the field I give you three examples why.

We all start with slime

You might be wondering why slime comes into a story about beautiful tropical coral reefs. Glistening aqua lagoons, gentle lapping of water on the shoreline, the sun shining through distant clouds as they gently drift towards the ocean bathed in the incandescent pink of a cooling summer's evening. These are the images that come to mind when people think about a coral reef. That scene of colour, wonder and beauty is an accurate portrayal of a gorgeous reef day from the angle that we (coral biologists) see it. As I said, coral reefs are everything, and I do get to work in spectacular places. I am not going to lie – being a coral reef scientist is living the dream. These places are without question spectacular, but just so you know, those glorious pink sunsets and aqua waters do have a few costs. For example, watching those sunsets while cleaning yourself after a long hot day in the sun with corals sweating slime or mucus all over you. Let's just say it's best not to stand downwind of a coral biologist at the end of summer! However, as a coral biologist, you will become just as much a fan of the mucus as you are of those island sunsets. An awful lot of slime

161

comes off hot corals as the sun beats down on the crystal-clear still waters, but corals are so much more than their slime. Let's start at the beginning and I'll show you why.

In becoming a marine scientist, marine biologist or coral biologist, no one really expects mucus to become a big part of their day, but when working in the field (on those gorgeous lagoons and islands) it is and we tend to find it very interesting. If you get up close and personal with these incredible animals – all of whom go by the simple name of 'corals', but are really a group of hundreds of different animals of all colours, shapes and sizes – you quickly come to realise that they are not simple at all. They are pioneers of the ocean, branching out wide (like corals in the Acroporidae family), stretching for the stars above the ocean surface, and creating homes all across this big blue planet that we share with them. Sometimes they live on their own (like solitary corals such as mushroom corals, known as Fungidae), sometimes in partnerships with others (like algae, known as Symbiondinaceae), sometimes they build enormous underwater cities teeming with life (coral reefs), sometimes they remain solitary and quiet, living in cold deeper places that us simple humans rarely think of as being home to corals. They do all of those things and more – and yes, it all involves slime. All of that mucus is an important part of their life cycle. It keeps them clean, provides food to other animals and, when the ocean heats up, it helps corals to survive the increased temperatures. So there you have it – my job is to understand coral, and working with corals starts with learning to love their slime. We begin these tales from the coral reef, not with shining blue lagoons and glimmering pink sunsets, but with slime.

I have spent many days, weeks, months and even years of my life studying corals. I know corals and their slime so well that one of the first things I notice when cruising into a reef, even before stepping off the boat, is the heady smell of that wonderful slime. Corals have a distinctive smell. It is a bit musty and sits heavy in the air. I always think of it as 'pink', a bit like the sickly sweet smell of musk sticks at lolly shops, mixed in with the briny smell of the ocean. It is a unique smell and as you get used to it, it becomes an inviting perfume. The smell of a coral reef is the first hint of all you are about to see and hear. It brings up memories of summer, the beach and lolly shops. I always smile when arriving on a reef because of that smell. I feel like I have come home for summer. I am horrified when anyone refers to these founding members of the most spectacular places on Earth as 'slimy rocks'. Next time you visit a reef, and before you step off that boat or into the water, face the breeze and breathe in the smell of the reef. Does it bring you memories of childhood summers, sunshine and lolly shops? If you happen to find me on a reef somewhere, I will eventually look up (keep reading and you will see why). Say 'Hi' and tell me how the corals are no longer just slimy rocks for you.

Moving to cooler waters

I have spent most of my years looking at corals on tropical coral reefs in warm waters, so travelling further south to cooler waters has been an interesting change and brought new experiences. I recommend visiting the reefs of cooler places; they have so much to offer. Norfolk Island is one of those places. The island is located far out into the Pacific Ocean, off Australia and a little towards New Zealand, but not as far as South America. It is a small but substantial island, and not one you want to miss. What Norfolk Island may lack in land size it makes up for in many other ways. It hosts one of the most spectacular and accessible coral reefs that I have ever seen, it has tales spanning lost civilisations and colonial life through to mutiny on the high seas, ancient Polynesian travellers who have left their mark on the island, and it is habitat to species that you will not see anywhere else on the planet – wonderful!

The corals of Norfolk Island are unlike most I have studied elsewhere. The ocean is cooler here (so corals are different from those of warmer waters), waves pound the island's reefs and the corals grow in extreme conditions. They are tough, unique and spectacular. To study corals, especially in a new place for the first time, we need to take time to observe them – examine how they grow, look at the other species that make up the reef, see which species interact with corals, and record the corals' population structure, health, growth and reproduction. All of these considerations are important in understanding how these places survive, thrive or change over time, especially if the environment is changing around them. They can't exactly get up and move away if the conditions change – but that is a different story for a different day. Watching corals requires quite a lot of floating, holding your breath, swimming slowly, being quiet, becoming part of the reef and its unique community.

Understandably, not all members of the reef like to be observed or watched (stalked, if we are being honest) and some reef-dwellers downright despise inquisitive strangers like coral biologists hanging around and staring intently into their homes. Who can blame them!? One such reef-dweller is the island's resident famous, feisty and fierce little damselfish – the banded scaleyfin damselfish, also known locally as the aatuti. I am not one to pay a whole lot of attention to fish – fish are not what I do – but this fish demands attention and is not to be messed with. The aatuti is a farming damselfish of the territorial kind. In fact, the species is fiercely territorial of its home. What it lacks in size – at least compared to the ferocious territorial human-hunting predators such as those shown in Hollywood-style horror movies (note: human-hunting marine animals don't exist) – the aatuti makes up for in determination to protect its part of the reef. The aatuti has no predators in this little lagoonal reef of Norfolk Island. It is the top-dog of the reef and it wields its power without remorse or fear. For a band of unsuspecting coral biologists staring intently at the beautiful corals, this fish poses quite the challenge.

Among all the coral counting, measuring, studying, watching, sampling, collecting, photographing, videotaping, more counting and more measuring that the coral biologists do on the reef, they also need to be on watch. Aatuti will be stalking them. They pop out from behind coral bommies and rocky platforms and before you know what's hit you – 'BAM' – an ankle bite followed by a yelp from a floating coral biologist. Laying our transects and setting quadrats to photograph, and out of nowhere – 'BAM' – another ankle bite and another yelp from a defenceless coral biologist. Swim around a corner of the reef, admiring the beautiful old coral bommie and the coral biologist suddenly ducks and weaves like their life depends upon it – 'BAM' – face to face with the ankle-biter. The coral biologist tries to swim-scurry away, but too late – 'BAM' – another yelp from the coral biologist. Who knew yelping would become part of the job? So, if you find yourself on Norfolk Island reef and see the normally slow-moving, manatee-like coral biologists covered up from head to toe, ducking, weaving and uttering the occasional yelp – you'll know why.

Despite their persistent hounding and nipping, we love these little fish and respect the fact that they are not ever going to love us back.

Close encounters and looking up

Coral reefs are not just made up of corals. No matter how closely you concentrate on the corals (and they are the best bit, really they are – I'm not biased at all), you will be distracted by other fascinating comings and goings on the reef. I have always expected fish to get in the way as I watch my beloved corals – dealing with pesky, inquisitive, interrupting and sometimes territorial fishes comes with the job. However, I've not had many close encounters of the shark kind. I am happy that they are out there; I just don't need to be up close and personal with these apex predators of the sea. Yet, on a reef things often have a way of finding you all by themselves. It could be the hours of floating, slowly moving along the reef, photographing, measuring and recording as the reef moves around you, or the many trips swimming back and forth, up and down, across the reef, doing all things science. I guess some encounters are inevitable.

It was a typically long hot smelly (remember all that slime and mucus) day on the reef. The last of the summer heat was beating down. It is heavy work swimming all day, working against changing currents, moving back and forth on the reefs, moving equipment around, tanks, seawater, setting up plumbing systems for aquaria, moving corals from reef to tanks, testing samples in the laboratory and more swimming. They are wonderful long hard-working days and by the time the hot sun starts to move towards the horizon, the day can be felt in your muscles and mind. One particular day, as the sun was falling, it was time for the last trip across the reef. Time for checking loggers and preparing for the tasks of the next day, washing the day off tired limbs and freeing ourselves from the cling of musty mucus smells. We always work in pairs in the water, helping and watching out for each other. We also always have another on look-out from the beach. The reef is a safe and wonderful place full of small beautiful things, but it also has the rare big thing. I have only once come across a big thing relatively close, and it was at the end of a hot and typically busy reef day. We had swum through the lagoon and along the reef edge much of the day (had done so for weeks, really) and knew it and its currents well. So, we were confident in having our heads down and getting on with the last jobs of the day. We paused to take in the last of the day from the vantage point of the crest of the reef. The ocean behind, the aqua blue lagoon in front and the corals below. This was the first time that day that we'd really looked up and lifted our heads above the amazing blue line that is the surface of the water, dividing the world below from the one above. What we saw above was definitely unexpected.

The aqua blue line of the lagoon was not peaceful. There was a thrashing in the water. We could see a tall black dorsal fin and some distance behind it was a smaller black tail fin belonging to one of the ocean's top dogs – a beautiful local reef hammerhead shark. It seemed to be having a snack. Hammerheads are currently being assessed for listing as a threatened species – so this giant had more cause to be concerned about us than we did about her, and we were very lucky to have this brief chance to see her in all her glory. One coral biologist went a little pale (let's be honest, a lot pale); the other said 'Wow, cool' and ducked his head under the blue line to try to see our new-found friend swim away. I was the one who went pale. They say that there are two kinds of people: I think that there are two kinds of coral biologists – ones that like to look up and ones that are happy being oblivious of the bigger inhabitants of the reef. Needless to say, the shark made its way gently across the lagoon (not so gently for the snack-sized ray it had eaten), passed us without a glance and swam over the reef crest back to its deeper ocean home. The hammerhead posed no threat to the tired coral biologists on the reef crest as it passed on its way, but it left a lesson – remember to look up. You never know what you might get to see and, importantly, when that might be your only chance to see it.

So where do these tales leave us? Well, they have left me feeling grateful that I have had the opportunity to know coral reefs, all the way from the smell of the mucus on hot summer days to the biggest creatures that call these places homes. I know that there are many more reefs to see, more adventures to come and wonders to find in the big blue planet we share with corals. In these tales I hope that you too find your way to a coral reef, maybe to many of them, and also find a world filled with wonder, beauty, mucus and the sounds of a busy coral reef.

The colours of our reef
Pink
Below the pine
Along the shore
Beneath the waves
they wait for more

Bright rising moon
Clearest night
A warm breeze rises
Oh, what a sight

Bundled up close
to be slowly released
All at once they go
to the surface reached

Dancing together
in the darkest night
the ocean turns pink
with touches of white

Below the pine
Along the shore
New life has arrived
More reef to explore

Blue
The sun sits low
The water cold
The reef's deep blue
heralds a legend of old

The powerful waves build
as winter arrives

but when ocean slows
its deepest secrets arise

Blue plates emerge
across the darkest of reefs
purples once hidden
now ready to seek

The blues of the reef
in winter they glow
when the ocean is dark
corals reach as algae slow

A treasure so rare
a reef so old
time has not forgot
their stories to be told

White
Summer arrives
Oceans warm in the sun
The days are long
Our Christmas is fun

Laughter and play
Sunburn and salt
Once again one with the reef
free, happy and bold

Days turn to weeks
not a whitecap near
no clouds to be seen
watch, with growing fear

All the fish swim away
departing the reef

all that can move
must escape this hard heat

First the reef sweats
shimmering alone in the sun
colours glow, fade to white
soon there are none

Ancient corals once stood
bearing witness to time
now slowly retreat
will they be lost to our kind?

Green
After white comes green
as the reef starts to change
once colour and life
now quiet and strange

They now see the reef slip away
where once childhood grew
they see their children miss
all that they knew

Now they see it is time
To learn from the past
go back to the old
we must change, new stories to be told

Return life as it was
all that swims will come back
we didn't wait for relief
we took the right track

Because first come the pinks
as they rush back to the shore

beautiful baby corals renew
then slowly grow more

You can see the purples, the blues
the secrets of old
Back to the reef we go
once again, so bright, so bold

Acknowledgements

Thank you to the wonderful women I have worked with, starting with Ruth Gates, who first inspired me and who I aim to live up to in everything I do; to those who kept me on the path and brought lots of laughs – Alana and Dana; and to the newest coral reef researchers who keep it fun – Char, Jesse, Shannon, Paige and Soph. Coral reefs need more wonderful women just like you all. Most of all, thanks to Bill, Marz, Pez and Ash – the rockstars I love.

23

People are strange: studying nature in cities

Dieter F. Hochuli

> What strange phenomena we find in a great city, all we need do is
> stroll about with our eyes open. Life swarms with innocent
> monsters. (Charles Baudelaire, *Les Fleurs du Mal*, 1857)

There's something special about being an urban ecologist. Admittedly, you don't get the same chances to showcase the natural beauty of some of the more spectacular parts of our planet, as ecologists who work in more remote places sometimes do. You need to make peace with the fact that you are often working on systems dominated by common species and numerous non-native species, which have significant negative impacts on more popular native animals and plants. Also, your field sites are characterised by something that many other ecologists don't have to deal with. People. Lots of people.

Doing ecological research around lots of people isn't necessarily a bad thing. We live in a rapidly urbanising world. The reality is that for a large part of the world's population, any engagement with nature happens in cities. Connections with nature are vital for the wellbeing of people, so from a scientific perspective it's urgent to understand how urban ecosystems work. However, I've learnt that funny things happen when you start trying to unravel the mysteries of how nature survives and thrives in cities. You notice the organisms you've been sharing your world with and start to look at them in different ways. You also see opportunities to share what you've learnt with a wider community that often doesn't know much about the amazing things they share their world with. It's probably taught me more than ever that we don't celebrate the extraordinary of the everyday nature we live with, and that when we do, it opens our eyes to a new way of thinking and talking about ecology and conservation.

***

It's no secret that people have modified the planet dramatically, often in ways that distress ecologists because of the impacts on the fauna and flora we hold dear. We call the current epoch the Anthropocene for a reason – human impacts on nature are everywhere.

In cities, modifications to the environment are extreme and seemingly so hostile that we often wonder how plants and animals can survive, let alone thrive, in these urban jungles. Yet they do, in surprising and innovative ways. So while the field sites often don't have iconic views that attract social media attention, they are home to some of the most remarkable stories of survival and adaptation in the natural world.

***

When you work in urban ecosystems, you will see some unusual sights. It turns out that greenspaces in cities, particularly those supporting vegetation with a slightly wilder and remote vibe, are ideal places for a range of activities including romance, mind-altering substance use and vandalism. The latter is often the bane of our experimental work when we deploy devices (like wildlife cameras and acoustic units) to monitor biodiversity, so we often search for places off well-worn paths in which to do our work. However, the reality is that despite our best efforts to bush-bash through a vast array of vegetation in surprisingly inaccessible landscapes, we rarely find places that are free from the signs of human activities. Given the bruises and scratches that we acquire in our futile searches for the relatively less-visited areas, you continually ask yourself 'What the hell were those people doing there?' Yet, when you reflect on their motivations, the evidence they leave behind and the human condition generally, you just shrug your shoulders and celebrate the diversity of behaviours that characterise our very complex species.

In terms of the evidence you encounter, it's typically quite unpleasant to spend much time dwelling on the histories behind it. We've found underwear in a variety of forms, sometimes in lonely singles and – perhaps a little more troublingly – in collections featuring a range of sizes and types. Burnt-out cars (occasionally still smouldering) also turn up in unusual places. We've found a range of bongs and pipes, homemade and commercial, littering the forest floor and coexisting happily with an extensive range of used condoms. Police tape turns up sometimes, making some of the sites unavailable for research in the short-term. All in all, not the easiest traits to market to prospective collaborators wanting to see some city-based nature.

Despite them being in plain sight, you become immune to a lot of these examples of human activities and learn to ignore them while you work. But this laid-back attitude can sometimes lead to awkwardness, particularly when you fail to warn co-workers about some of the things they may encounter. One of our long-term field sites in Sydney Harbour National Park has several remarkably secluded beaches that are enjoyed by naturists throughout the year. I regularly visit these sites to work on caterpillars that show boom and bust cycles, feeding on the dominant eucalypts and angophoras in the area. I take interns and summer students to assist with the work as part of their research training. The work requires a focus on the caterpillars and the leaves they feed on, to the point where it is very easy to stumble into a space where older gentlemen sporting dark brown leathery skin are sunning themselves in all their naked glory. While these incidents haven't proven to be a problem of any significance, the failure to warn aspiring young scientists that they may encounter these gents has resulted in gasps of surprise on multiple occasions, followed by an apologetic 'Oh yeah, there are quite a lot of naked older men at this site'. Admittedly, the sign erected by the national parks authority that points to the beach and reads 'Nudity permitted on beach only' should have reminded me that my risk assessments ought to include the possibility of encountering platoons of naked sun-worshippers.

***

Most of the science around plants and animals in cities is driven by fundamental questions like 'Why are you living there and how do you do it?' As such, the techniques we use to study the natural world in cities aren't really that different from methods used in any of the systems that ecologists frequent. We observe, collect, use some idiosyncratic pieces of equipment, and wander around places that our plants and animals like to live in. The challenge for many of us is that, to dozens of people, what might be perfectly normal behaviour for an ecologist looks awfully suspicious to the broader community.

Take studying pollinators as an example. One of our most basic methods for understanding pollinators involves simply stopping and staring at flowers for a set time, noting the activity of floral visitors, then moving on to the next group of flowers. Innocuous enough? It turns out that it's a fine line between studiously watching pollinators and loitering in a manner that attracts the attention of eagle-eyed citizens. You inevitably get used to speaking with police about your intentions and your motivations for wandering around the urban forest. It does lead to amusing conversations about why your job requires you to lurk through the bushes, staring intently at flowers! Some of these chats end with participants having a newfound interest in the natural world. Others end with awkward exits when people realise exactly what your day job entails and whose taxes are paying for it.

While wandering among the bushes can raise suspicions, I've found that some of our equipment is the perfect conversation-starter. If you are seeking attention from the wider public, no better accessory exists than a butterfly net. More than any other gear, the butterfly net seems to identify us as mostly harmless (a counter-intuitive notion to any readers of John Fowles' 1963 novel *The Collector*) and invites curious minds of all ages to come over, watch what we're doing and ask why we are doing it. Indeed, the great gift for ecologists who work in urban ecosystems is that the people who are central to creating the systems are curious about how the systems work and who else lives in them. The one massive lesson I've learnt from all the conversations with people in the field is that, as an urban ecologist, I have an obligation to share what I know about the natural world with the diverse group of people who live in cities. The rewards for doing so are immense. They have spurred me on to try and get better at telling our stories and finding new ways to engage with different people.

***

One of the things you can't control when doing ecology around lots of people is that when you encounter the curious ones, the types of questions they ask will be wide-ranging, often out of left field and regularly on subject matter that might not be in your disciplinary wheelhouse. These situations require a decision: do you acknowledge the imposter syndrome that often comes with academic life and defer the questions to more knowledgeable people, or choose the path of becoming an instant expert in stuff ... Lots of stuff. Choosing the latter path may seem a bit intellectually arrogant given the extraordinary knowledge base of disciplinary specialists, but working as an ecologist in cities has taught me that it's important to embrace the benefits of being an ecological generalist. This realisation is anathema to much of the career advice given to early career ecologists, but over time I've come to see the benefits of challenging yourself to learn about all the things that you don't know much about – and then trying to get better at talking about them to different audiences.

Regularly being flummoxed by the identities and behaviours of animals and plants can be powerful motivations to learn more about the organisms that form part of your direct expertise. The science we do still reinforces the merits of specialising on certain groups or systems. We wouldn't be able to ask and answer detailed questions about the natural world without specialising, but for me the interactions outside of academia have reinforced the importance of finding time to develop authentic generalist skills and diversity in thinking. It's always surprised me that as senior ecologists, despite being a group of people who extol the importance of diversity in natural systems for the maintenance of life on our planet, we encourage people to specialise more and more in their academic journey rather than embrace general and diverse skillsets. While it's important to overcome your imposter syndrome, be a generalist and embrace your inner instant expert, it's also easy to become a little too confident. It's one thing to be able to provide a plausible answer to questions like 'What is that?' or 'Why is it doing that?', but as you engage with people in different environments you will encounter rather brutal reality checks.

I've enjoyed taking interested participants on 'walkshops' to explore their local environment. One of my more humbling moments happened at the start of one such event when I decided to share some of my (then) newfound bird expertise, by pointing to a pair of tawny frogmouths in the distance. These much-loved birds had adopted the signature pose of their species, perching on a branch with their heads pointing skywards. Confidently sharing numerous factoids, including about how these nocturnal marvels were more closely related to nightjars than to owls and were voracious consumers of the invertebrates that many of my audience perceived as pests, I took the group right under the tree. The 'birds' turned out to be a fractured tree branch, which, in my defence, had two prominent knobs of wood that looked remarkably like a pair of perching frogmouths. This incident was not one of my more auspicious forays into science communication.

Mistaken identity can work both ways. One of the great joys of working in urban systems around Sydney is seeing the remarkable serpents that persist in remnant vegetation in the city. I vividly remember stumbling across a red-bellied black snake while leading a group on an urban nature walk. After the brief shock that comes from encountering a snake, I set out to make it a memorable and safe experience. Informing the group of the snake's presence and organising everyone to move slowly so as not to startle the animal that was basking (in hindsight, in unseasonably cold weather), we slowly approached ... some black rubber tubing coiled in the grass, left after a recent planting effort. Unsurprisingly, the tubing neither slithered away nor showed any interest in biting us. Less than a week later, I took a (thankfully different) group to the site for a similar walk. As I blundered towards an old tree that was home to multiple nesting birds, another carelessly discarded black rubber tube less than a metre away slithered off rather grumpily into the grass, with a distinctive flash of red on the underside of its glorious black body. I guess both situations were teachable moments, but not in the way I'd really intended them to be.

***

My personal outlook on research, teaching and outreach in ecology has been profoundly shaped by the people I've encountered in the field and by the oddball collection of animals that make their homes in cities. The people part took a bit of getting used to. People are embedded in these systems and their opinions, fears and interests are all part of understanding how nature in cities can be conserved and promoted.

Human–wildlife conflict is a real challenge to urban ecologists, and channelling our enthusiasm for species like snakes, spiders and even fewer polarising species is one of the real frontiers for helping promote connections with nature. One of the great revolutions in ecology has been the rise of citizen science and the embracing of enthusiastic non-specialists in helping us understand the natural world. The observations and data provided by these extra sets of eyes have been helped us ask questions that we would never have been able to address with our own limited resources, such as how different species modify their behaviours in response to the challenges of living in the novel ecosystems they inhabit. Participating in the process of science, rather than being a passive receiver of science communication, provides a great opportunity for science to be more diverse and inclusive.

I've also developed a real passion for understanding the ecology of the lesser loved species in our urban ecosystems. Whether it's brush-turkeys recolonising their former homes and remodelling suburban parks and gardens, or hover flies doing a lot of the heavy lifting in pollinating (compared to our native bees!), observing these animals thriving in cities and undertaking long-term research on them has been a gateway towards sharing their stories more widely. This research shows how these animals make their way through the world, and gives us the narratives we can share to wider audiences. It doesn't necessarily win everyone's heart and mind, but seeing someone begrudgingly acknowledge the work ethic of the Australian brush-turkey while their garden disappears into a nest mound, or proudly tell their friends that flies are just as important as bees when it comes to pollination (and are wholly responsible for chocolate), is a wonderful sign that a little bit of knowledge and understanding can go a long way towards making us a bit more tolerant about some of the animals we live with.

***

Aspiring and experienced ecologists are typically captivated by far-flung places on this extraordinary planet and by the amazing animals and plants we share it with. I'm no different, but my work in cities has made me realise that some of the things we take for granted, living beneath our noses, have remarkable stories to tell. More importantly, we still know so little about so much when it comes to these animals and plants. Whether it's the 'out of sight and out of mind' insects and spiders that are going about their daily lives, or the abundant species that are a nuisance to some, the responses of the natural world to the selection pressures imposed by the hostile environments of cities is one of the great opportunities for ecologists. Understanding them helps us tell their stories and find a way for city-dwellers to engage with them, making the nature connections that are central to the wellbeing of residents and the sustainability of cities. I've found enormous inspiration in everyday places and common things while wandering around bushland in cities. Some of the most beautiful things I've seen in nature have been in the heart of big cities. If I have one message, it's to take the time to look at and see the extraordinary in the ordinary. It's all around us.

Acknowledgements

I'd like to thank the many people who I have spent time with in the field. I'm grateful for the patience and good humour of the postgraduate and honours students who have taught me so much as they started their research journeys, and the collaborators who have generously shared their expertise with me and indulged my wide interests. I'd also like to thank everyone who asked questions about what we were doing. Whether it was the sceptical stare when you weren't convinced or the sparkle in the eye when you were, it inspired me to learn more about the many things I was ignorant about, or to do a better job of explaining why we do our work.

24

A dingo gold mine

Bradley P. Smith

For scientists who enjoy creature comforts when they are in the field, I may have found the perfect place. About 500 kilometres east of Port Hedland, in Western Australia's Pilbara region, lies a 20 square kilometre hole in the ground. As you fly across the endless rows of sandy red dunes where it is located, the hole reminds me of a dilated pore on a human's face. A scar or blemish for sure, but one that is dwarfed by the vastness of the Great Sandy Desert. The owner and creator of this giant hole was an international gold mining company that contracted me to help it manage the resident dingo population, some of whose behaviours conflicted with the mine site's operations and zero tolerance of safety risks for staff.

One of the mine's environment officers emailed me out of the blue, wanting to share her experiences of working with these dingoes and her observations of their Houdini-like abilities. As a research psychologist who focuses on dingo cognition, my profile came up on Google after she searched for 'dingo intelligence'. I was eager to see examples of dingo behaviour in wild settings, so after my enthusiastic response, she presented me with photo and video evidence of dingoes escaping from cage traps. Some of their escape strategies relied on brute strength, but in some cases they had to rely on pure wit and cunning. Dingo-related behaviour research findings generally attract global attention, but, locally, little funding or opportunities exist to carry out the work. Here was a prime opportunity to change this. Despite having only recently completed my PhD and having zero experience dealing with industry for project or consultancy work, I optimistically asked if I might come to see the dingoes and offered to help with the company's dingo management plan. It worked, and within six months I was heading west for the first of several field trips. I was keen to learn more about free-ranging dingoes (till then, my experiments had been conducted only on captive dingoes), but I was more excited to have found a way of using my background and experience to address an applied problem – trying to reduce the negative impact of dingoes on the mining operations.

Dingoes are native Australian canines found across the mainland. They are often referred to as the 'king of the Australian bush', given that they are Australia's largest mammalian predator (other than humans) and are at the top of the food chain. Yet, on average, they only weigh around 16 kilograms and are similar in size to a North American coyote. Much like coyotes, dingoes are highly adaptable to changing conditions, able to occupy all habitat types (from deserts to the alps) and kill and eat a variety of prey. In all ways, the dingo lives the life of a wild canid, with no reliance on humans at all to survive. Unfortunately, because of its threat to livestock, the dingo is often considered a pest across much of its range and is subject to constant lethal control programs.

You might imagine that studying dingoes in the desert involves long days driving and trekking through red sand in search of signs of their activity (tracks and scats), sleeping in tents and lots of sweat. Ordinarily it does, but not this time. The mine was a self-contained mini city, complete with airport, air-conditioning, 4G phone service and a full-service food hall. Everything was giant except the dingoes, which had come from far and wide to make a home in this mining town. Finding dingoes each day was not difficult. I often came across one on my 6:00 am walk to breakfast from my donga (portable accommodation). Mines were a resource-rich place for dingoes, as they provided year-round food and water, and structures (roads, pipes, tyres, topsoil dumps) that provided the necessities of dingo life, in return for little effort beyond dodging what the miners call light (utilities and four-wheel drives) and heavy (road graders, dump trucks) vehicles.

However, the mine happened to be in one of the hottest places on Earth – the average temperature in summer was a brutal 39°C. These conditions epitomise the definition of 'dog days' – an Australian colloquialism that describes weather so devastatingly hot that even dogs can do nothing but lie around on the baking earth. Observing how the dingoes dealt with the heat, living in a wild setting with abundant resources surrounded by busy people in high-visibility clothing, was an insightful experience. Indeed, many aspects of intimate dingo life are unknown or unverified, mostly because of the difficulty in locating populations and getting close to wild dingoes. Visiting the mine afforded me new insights into dingo life generally, including how they behave during flush times (when resources are plentiful), and what happens when you stick a bunch of adult humans in a small makeshift town in the middle of the desert.

At their peak, the dingo population swelled to around 120 individuals across the mine's footprint, which consisted of the operational area (open pit, offices, waste rock dumps, cyanide ponds, topsoil dumps, offices, mechanics and processing plants) and a village supporting up to 400 human residents, located about 20 minutes away (with restaurant, airport, accommodation, pool and sports facilities). Given that the dingo population in similar desert environments would consist of a small family group of four to six individuals per 100 square kilometres or more, for dingoes, this scenario was the equivalent of a starry-eyed teenager moving from a small town like Broken Hill to the bustling metropolis of central Sydney.

Unlike feral domestic dogs in a similar scenario, the dingo sub-groups maintained distinct territories (around six), preserving strict borders and denning in unique areas – much like North American wolves do. Some areas of passage were used by all groups, particularly around roads and the rubbish tip, but overall the separate groups seemed to cohabit the mine with a degree of order. Of course, every so often a fight broke out as boundaries and tolerances were tested. Tracking of individuals from neighbour-adjacent packs, using collars fitted with GPS, revealed that the rubbish tip had been arbitrarily divided right down the middle (akin to east- and west-side gangs). No dingo ever ventured across the midline in the six months they were monitored, despite the absence of a physical barrier that would prevent them from doing so!

The density of dingo populations is directly related to the availability of resources critical to their survival (food, water, and shade). The rubbish tip gave access to food scraps, and the many ponds –whose water had low levels of cyanide that filtered through waste rock to extract any last remnants of gold – provided adequate year-round water. The dingoes had quickly eradicated the available natural prey, which had no chance against so many skilled hunters. Analysing their scats revealed junk food diets, but also that the dingoes had become cannibals. The dingo death rate was high, perhaps because of vehicle strikes, ingesting water from the cyanide ponds and various other chemicals found on industrial sites, as well as fights between dingoes. The animals certainly took advantage of this, scavenging the carcasses and leaving no trace except the skull.

Normally dingoes are crepuscular (active at dawn and dusk) to avoid the heat of the day, and to coincide with the activity of their prey. Yet, the dingoes here were active throughout the night and day. This difference was likely reflective of the free time they gained, because they did not have to spend hours finding food or in rest and recovery. I had an air-conditioned vehicle to which I could retreat every few hours but keeping cool remained a major challenge for the dingoes. Again, the mine site conveniently offered various ways for them to cope. During the day, you could find dingoes taking a dip in the ponds, lolling under the shade of dongas and buildings, hiding underneath stockpiles of giant truck tyres and various infrastructure, or playing in the water dripping from leaking pipes. At one work site, the dingoes knew that 4:00 pm was when the vehicles were washed to rid them of the red dirt, and they came to lie in the cool runoff. The dingoes would also dig out resting pads, like small caves, seeking cooler earth inside dirt mounds and in the sides of creek beds to rest.

One day, the environment officer and I were walking around a topsoil dump area with its large mounds of rich topsoil, scraped from the surface and stored in piles. The stored topsoil is used later when the mine replaces and rehabilitates the ground disturbed when mining. These mounds provided the perfect habitat for dingo dens. We first noted signs of denning in the side walls of these mounds, with clear scratch marks made by dingo paws. These marks turned out to be failed, or early, attempts at den building. After properly surveying the area, we discovered 24 dens in an area the size of an Aussie Rules football oval. Four of them were active, housing pups. We measured the dens (depth, width, orientation of the opening, vegetation in and around the den) and learnt much about how dingoes build dens. One den, positioned on an idyllic site overlooking an open area, contained a surprising number of puppies – 18 little balls of fur, much larger than the average litter size of five. The different sizes of the pups made it clear that at least three litters were present, suggesting that several mothers had banded together to co-parent and form a creche for their pups. This observation contrasted with existing theories that dingoes regulated populations by dominant female infanticide – a senior breeding female kills the pups of any other breeding females to increase the survival rate of her pups.

Excited by this discovery, but keen to limit any disturbance to the pups, I placed a GoPro on a tree overlooking the den and promptly left. The next day when I went to retrieve it, it had gone! The only way they could have retrieved it was by jumping about 2 metres in the air and pulling it off the branch. The loss of equipment didn't concern me, but the images inside were irreplaceable. It took two days of searching, but I recovered the camera. Teeth marks were on the waterproof plastic shell but the innards were unscathed, and I got some amazing footage of the mother returning to the den after we had left (no adult dingo had been present when we were there) and the pups coming out to greet her. The footage was exhilarating! I had never seen such behaviours before. I was giddy with excitement, both for being able to record the moment and for gaining a glimpse into their secret lives.

The day that GoPro was found, I again approached the den to count the pups and take photos for evidence and later analysis. I knew that disturbing dens could lead dingo mothers to relocate pups (or worse, abandon them), so I was careful not to get too close or disturb the area. The following day, I was horrified to find that the den was unoccupied. The adult female had likely just moved the pups to a nearby den, but I felt terrible that I might have stressed the parents and caused this response. I learnt a valuable lesson that day about limiting my impact on wildlife. The surveys led to the first published insights into the denning and whelping behaviour of dingoes, but data should not come at the expense of the wellbeing of the animals you are trying to learn about and help conserve.

The discovery of these dens led to another significant interaction I will never forget. On one of the afternoons spent surveying the denning area, I noticed I was being followed. I didn't realise how many dingoes were in pursuit until I found a relatively open area among the spinifex and acacia trees, where I knelt to do a count and spend some time observing them. Six males approached within 2–3 metres and surrounded me. The two in front had their hackles up (a sign of uncertainty but also aggression) and began growling, barking and leaping towards me with their backsides in the air (similar to a classic domestic dog play-bow), while a younger dingo started playfully digging in the sand and playing with a nearby stick. While my attention was on the dingoes in front of me, a dingo quietly crept towards me from behind and attempted to pull away my backpack, which I had placed on the ground, with his teeth. It's always the ones behind that you must watch! It reminded me of a scene from my favourite childhood movie (and early inspiration for studying predators), 'Jurassic Park'. In one classic scene, Robert Muldoon, the game warden, was tracking the escaped velociraptors but realised too late that he had been outsmarted and they'd been hunting him.

However, unlike Muldoon I made it out unscathed. I knew what to do (or perhaps what not to do) around dingoes so as not to escalate the situation. Still, I'll admit that I was nervous at one point – especially in somewhat unfamiliar surroundings so close to active dens. I knew that I could be in trouble if the situation escalated. I remained in a kneeling position taking video and photographs, keeping still and non-threatening. After five to 10 minutes of this intense standoff, the dingoes seemed to relax, and most either lay down to watch me or went to sleep. About 10 minutes later, I decided it was a good opportunity to leave. They walked with me, escorting me back towards the dirt road like a private security detail. It reiterated to me the importance of safety in numbers when out in the field (a minimum of two people). I also took comfort in the knowledge that dingoes, no matter where I have interacted with them, act predictably. I feel safe around them, knowing that they do not see adult humans as prey and that their actions appear to be based on curiosity more than aggression – even when they have young nearby. Dingoes appear to spend most of their time assessing how people react to them, almost like a test in which they are determining whether we are prey, playful or a threat. To me, dingoes seem far more predictable than domestic dogs – where any breed of dog or individual can act differently despite having the same experiences.

Being surrounded by a pack of dingoes gave me insight and empathy towards the miners who had been approached or surrounded by dingoes while trying to perform their duties or in their free time. It can be an intimidating experience. However, the dingoes' responses also highlighted that where negative incidents had occurred, the situation had likely been unwittingly escalated by the person's reaction and response. Such incidents led to questions about risk – a topic high on mining companies' agendas. I was asked to assess the level of risk dingoes posed to the miners. I concluded that the risk to people from dingoes was low to negligible – provided nothing major changed at the mine, and people acted with common sense and in accordance with existing safety recommendations. This opinion was based on my own interactions and observations of the dingoes, other people's hundreds of daily interactions each week over many years, the lack of children and pets on site (typical targets), the lack of frequent historical dingo attacks not only locally but in Australia more broadly, and analysis of the mine's incident report register.

Unfortunately, risk mitigation is simply that: mitigation. Risks in complex environments are often difficult to eliminate, particularly when the behaviour of free-willed humans and wild animals influences the outcome. A few years after my research concluded, a person was mauled by two dingoes at the mine while on their lunch break. I was called to be an expert witness in the proceeding court case. The outcomes of the legal proceedings are documented elsewhere, but a key learning from this situation as a researcher was that documented records, including reports and consultations that you produce, should be written as concisely, objectively and unambiguously as possible for these situations.

The relationship between people and dingoes is complex, but perhaps neatly summed up by a short encounter that occurred on site. Adam (not his real name), a manager at the village, watched as a bloke arrived at his room with a pizza and six-pack of beer. The miner placed the pizza box on a chair next to the door, and carefully rested the beers on top of the box. He fumbled in his pockets to locate the keys to his door. After opening the door, he turned towards his dinner – just in time to witness a dingo pulling the pizza box off the chair. In the process, the beers tipped over, exploding as they crashed onto the concrete floor. The dingo, seemingly unfazed by the foaming mess, trotted off with pizza. Adam, laughing hysterically, yelled out in sympathy to his colleague 'Sorry mate, that sucks!' The miner, relocking the door behind him, muttered in frustration 'I fucking hate dingoes', and began walking back to the take-away shop to replace the stolen pizza and spilt beer.

Stalking the miner and waiting for the opportunity to take the pizza, highlighted the dingoes' adaptability and intelligence. It also exemplified the varied nature of people's relationship with dingoes. Some people love them, some hate them and some just tolerate them. The mining executives (who did not live on site) had little tolerance for dingo 'risk' and even ordered three dingo culling events during my time there. However, generally, dingoes were welcome. In interviews I conducted, employees described joy in observing wildlife in an otherwise sterile industrial environment. Especially when dingoes were engaging in wild behaviours (hunting, playing, howling) rather than being destructive or disruptive (stealing food, but also bags and equipment). The environmental officers had instigated many positive strategies to try and manage the dingoes and limit their more disruptive tendencies; for example, erecting fences, installing locks on bins, and using light globes that did not attract moths – dingoes were often seen hunting moths around artificial lights in the evenings. Dingoes were an accepted part of the mine, and most people appreciated that they were part of the local desert environment.

In the end, this field site was a literal gold mine in terms of learning about dingoes and the ease with which they could be studied. At my current field site, only a few hundred kilometres to the north of this mine, seeing a dingo is a rarity. After three trips, I have yet to witness a dingo in the flesh – instead, I rely on what is recorded on motion detection cameras. Perhaps they have all moved to a nearby mine?

Acknowledgements

The research was undertaken on traditional lands of the Martu people. I wish to thank Anne-Louise Vague for making the research possible, and Mia Cobb, Rob Appleby and Helen Waudby for providing helpful feedback on early versions of the chapter.

25

Bogged in the desert

Helen P. Waudby

Thunder and lightning lasted most of the night; caravan shook.
Worried about being struck. Put on thongs. Port nearly finished.
So much for any port in a storm. (Field notebook entry,
September 2009)

It is 4:00 pm at my desert field site and a dust storm is raging – the third
this week. Typically, the dust precedes thunder and lightning shows
that sound like they are rolling across the desert. Sunlight filters
through the fine particles, casting an unsettling orange glow. In a few
hours this wall of dust will reach Sydney (2,000 kilometres east of my
location), covering that major Australian city in the same eerie light. I'm
protected by the thin but sturdy walls of an ageing 1970s caravan. I open
the door a little and peer out into the raging storm. A sudden gust
wrenches the door from me like an invisible hand and flings it open.
Sand pelts me in the face, flying up my nose, and into my eyes and ears.
As the distant rumbling starts, I sigh heavily. Not for the first time, I am
struck by the irony of working in a desert only to be deluged by constant
rain storms.

***

I started my PhD research in 2009 near the end of Australia's Millennium
Drought, on a 4,000 square kilometre cattle station just south of Kati
Thanda-Lake Eyre, the traditional lands of the Arabana people, in
Australia's arid interior. I was travelling to this site (~10 hours drive north
of South Australia's capital city, Adelaide) to try and understand how
cattle grazing affected the native animals and plants that lived in this arid
and fragile environment. I was prepared for hot and dry conditions,

given that the region had barely seen rain for the past decade. Not long after I started my research, an intense La Niña weather phase settled over Australia, bringing with it intense rainfall. As a result, the parched desert became lush and green, and small desert mammals that were rarely seen in that area, like desert mice and plains rats, and introduced house mice, became relatively abundant with all the extra food. However, the rain also made access and travel to study sites difficult, particularly as cracking-clay soils and gilgais (depressions in the ground created by the shrinking and swelling of clay minerals that fill when it rains) were a key feature of the local landscape.

Bogging my four-wheel drive vehicle became a regular pastime. My usual recovery technique involved shoving ready-made plastic 'tracks' under the tyres, reducing tyre pressure and unloading the vehicle as much as I could. As I worked mostly alone in a remote area, I needed to be relatively self-sufficient which meant that I carried the proverbial 'kitchen sink' level of equipment with me. Consequently, the process of unpacking was extremely tedious. The vehicle's winch was generally unhelpful in these situations because the cracking-clay plains were largely treeless (no points to winch off). I became a proficient de-bogger, but some incidents still managed to get the better of me.

Early in my research, I learnt an important lesson about the consequences of driving on wet roads in the desert. It was my first major field trip, and one of the few where I had company. Jane, Mike and I had camped in a wide, sandy coolabah tree-rimmed creek. We had planned to visit the town of Roxby Downs mid-way through the trip to stock up on essentials like beer and lettuce. The town was established in the mid-1980s on the traditional lands of the Kokatha people, to service the local Olympic Dam Mine, and was about two hours' drive from the station – when road conditions were good.

My map of the station indicated that we could access a road that exited through the north of the property and onto the Oodnadatta Track, from where we could travel to Roxby Downs. The evening before, torrential rain had fallen during an impressive lightning show that sent us scurrying for the car, where we sat arguing over who should run through the storm to turn off the diesel generator powering the caravan (I was nominated, as trip leader). The next morning, naively admiring the glassy and extensive pools of water surrounding us and navigating increasingly slippery roads, we travelled slowly along the northern road, which turned out to be little more than a goat track. We were told later that it had not been driven on in years. After several hours, we ended up on the Oodnadatta Track, which wasn't sealed but was at least a proper road. The rain started again, pelting the car roof and creating a somewhat anxious atmosphere – the sound of rain on a car roof still makes me tense to this day. On reaching the outskirts of Roxby Downs we passed a yellow ROAD CLOSED sign. Having not had internet access for the past week, I did not know that most roads in the area were closed. It was clear that we would not be returning to the station by that route.

We bought supplies and headed to an unsealed road south of town, which passed through several local stations, to see if it was passable. Growing up on a cattle station in the central Australian desert had taught me how treacherous unsealed bush roads can be after rain, and I had – or should have had – a solid knowledge basis. However, optimism can make you foolhardy, and as no road closure signs were in place – because it was not a council road, as we realised later – we merrily set off on the drive back to camp.

We dodged or drove through stretches of road that had become lakes as it continued to rain steadily. The road eventually disappeared under a vast expanse of water, extending over 500 metres. It could just be seen re-emerging in the distance, if we squinted. I evaluated our chances. If the track was too soft in the middle and we got stuck it would be a very wet and uncomfortable process to extricate ourselves. So I decided to drive off the road, to avoid the water and potentially boggy track. Dear readers, NEVER leave the road. It had not occurred to me that the road was likely to have a more hardened base than the soggy clay country surrounding it. It also did not occur to me to walk the path in front of the car first.

We became very bogged, very quickly, as the car sank through the soil crust. With frantic digging, piling of rocks under tyres (I did not have my magic plastic tracks at that time) and swearing, we freed the vehicle. Our elation manifested as wild dancing (by me) and air guitar playing (by Mike) as Jane drove free of the sucking clay. We had gotten off relatively lightly. We agreed to return to Roxby Downs, where we would wait for the roads to dry a little. Unsurprisingly, since the rain had not stopped, the pools of water behind us had expanded even further.

Around 2 kilometres from the safety of the bitumen, we reached another particularly long stretch of water. I clearly had not learnt from my earlier experience and decided to drive around it, again. The car sank immediately and heavily; my stomach dropped with it. I spun the wheels, no doubt assisting the car to settle more deeply into the mud. The three of us sat silently, staring out the windscreen into the rain. I sighed heavily and opened the door to assess the situation. As indicated by the sinking sensation and lack of forward motion, we were bogged to the axles – the tops of the tyres were barely visible. Unwilling to admit defeat immediately, we dug for a couple of hours, placing rocks and vegetation wherever we could dig a gap under the tyres. We tried to maintain good humour by singing. 'Highway to Hell' seemed appropriate.

Unfortunately, no amount of digging or singing would budge the car. Evening was settling, and we were cold and covered in mud. With our only other option being an uncomfortable night in the car, we decided to walk back to town. We packed essentials, left a note with my phone number on the dashboard, and started off with an unjustified level of good cheer. We walked the remainder of the track, which was annoyingly firm underfoot, reaching the bitumen in the pitch black. Roxby Downs was another 20 kilometres north of our location, and none of us was enthused about the long walk to town.

I was equally unenthusiastic about hitch-hiking when Jane suggested it, but without mobile phone reception or a satellite phone we were out of options, so we waited. A white ute drove past, and we attempted to flag it down. No luck. We kept walking, feeling rejected. Minutes later the same vehicle drove back to us and stopped. The driver opened his window and said 'I was going to keep going, but there's no one else behind me and you might have trouble getting a lift in your state anyway'. With great relief and gratitude, we crammed ourselves and our gear into the car, trying (unsuccessfully) not to muddy the upholstery. Our saviour turned out to be a worker at the Olympic Dam mine. He was bemused by the explanation of our predicament and dropped us at the local caravan park where we stayed for the night.

The next day I made an apologetic phone call to the owner of the station where I was doing my research. She was concerned (and annoyed) because they had checked the camp and we were nowhere to be found! I told her that we were trying to find a local vehicle recovery company to drag our car out of the mud and that we would drop by the homestead when we returned. Her parting response was 'Good luck'. I rang the one or two local companies. Neither was enthusiastic. One commented, 'We don't go out on those roads in this sort of weather', to which I thought 'Yes, thanks for the tip mate ...'

I was fairly despondent when I answered my ringing phone but perked up quickly. A local station manager had found the sunken car on their property, with my note on the dashboard. I explained the situation and asked tentatively for help. Few words were exchanged, but she gave me hope, saying that she'd speak to her husband, Bill. Bill rang me back and I reiterated my predicament, probably rather plaintively. Bill, who appeared to be a man of few words, was clearly mulling over the situation. After five seconds of silence, he said 'I might be able to help.' I was very thankful, and even more so when he said that he'd pick us up from the bitumen.

Bill met us at the bitumen as promised, and when we arrived at our car we were surprised to see that its wheels were visible – when we left it, they had been buried completely in the mud. As I took in the situation, I realised that a not insignificant hole sat behind the car. Bill nonchalantly commented 'looks as if someone already pulled yer out'. He had dragged the car out of the muck before picking us up – probably thinking (fairly, based on our predicament) that it would be easier without us in the way. Despite our offers, Bill would not take any compensation for his help. In the end, we thanked him profusely for the 50th time, got into the bedraggled car and picked our way carefully back to the bitumen.

The roads reopened the next day and we were able to travel back to the station on the same goat track we'd left on. The track had become extraordinarily slippery and we slid off it a few times, which was a little hair-raising but not such a problem at 30 kilometres/hour. At the homestead I sheepishly apologised to Col and Jill, the station owners, for leaving the property without telling them. I probably appeared suitably chagrined and downtrodden. Fortunately, they had a sense of humour (they are good friends still) and I think were amused at my naivety and my downtrodden demeanour. I was assured that Santa Claus was not coming for me that year. Oh well. I was far more cautious about wet weather driving from then on, and the car was later fitted with better tyres.

***

I continue reading, trying to ignore the rain and my unease. I am quite exposed out on the treeless plain in my metal box, probably feeling much like a dunnart (a small carnivorous marsupial) in one of my metal traps. Thunder and lightning continue. The wind howls and rocks the caravan slightly, leaving me cold and a little sweaty with nervousness. I am sure that either the creeks will rise and the caravan (and I) will float away in floodwaters rivalling those of the 1970s, the wind will knock the caravan over or, worst of all, the caravan will be struck by lightning. While I hope that any lightning strike will target the metal windmill 200 metres from the caravan instead, I put on my rubber thongs (shoes) and some pants in case it doesn't. Maybe the thongs will insulate me against a lightning strike? If they don't, at least the donning of pants means that whoever finds my charred remains won't be embarrassed by my semi-nudity. I lie awake all night picturing each scenario – the storm reaches its crescendo somewhere around midnight.

***

On another occasion, my traps were open when an unexpected storm rolled in. I awoke around midnight to the sound of rain lightly smattering the caravan's roof. I leapt out of bed (my feet getting caught in my swag), grabbed my boots and started pulling them on while stumbling towards the car. My pitfall traps (basically buckets or tubes buried level with the ground) had to be closed before the rain really set in. Failure to do so could mean that any occupants of the pitfall traps could drown. Even worse, I could be cut off from the traps by flooded roads and they would be left open and unchecked, potentially trapping animals for days. I drove up the road while lightning flashed in the north-east – right over my trap lines.

Upon reaching the site, I ran along the traps, checking and closing them hurriedly. Despite my haste and concern about getting marooned at the site, I was overjoyed to find a single sleek and rather plump planigale (another type of small carnivorous marsupial) peering up at me from the bottom of the trap. Planigales are among Australia's tiniest animals. I didn't catch these fierce little desert-dwellers often, as they were quite cryptic, mostly living among the deep soil cracks. I retrieved him from the trap, oblivious to the advancing lightning for a few seconds while I admired his shiny coat and wedge-shaped head, perfectly shaped for living in the soil cracks. Lightning suddenly flashed in the distance, startling me into action. I released the chubby planigale near a crack in the soil (he dove straight in), and I recommenced closing traps. I arrived back at camp just as it really started to pour. The rain was torrential and thunder had joined the lightning, which was illuminating the desert.

A year earlier, I would have been huddled in the caravan, wishing the storm away. But this time all I could think of were the flocks of neon-green budgerigars that will descend on the seeding grasses that spring to life because of the rain, in the months to come. I also thought of my late father, a station owner, and his love of the rain. I recall that during rare big rainfall events on our station, dad would pace around the homestead veranda occasionally letting out a 'WHOOP' or a 'YOU BEAUTY' at the top of his lungs as the sky thundered or lit up. It was amusing for my brother and I, who were used to seeing dad behave more sedately. For him, rain was the source of life and income. Because of the rain, we could grow the green feed that would turn cattle fat over the next few months and attract all the grassland birds that he loved to watch.

In the morning, the storm had passed and I was greeted by a calm blue sky mirrored in the gilgais surrounding camp, which had filled with water. It was a big downpour and the creek was flowing. Days would pass before I could leave camp to reopen traps, but I didn't mind. I sat on the doorstep, watching the galahs arguing over seeds and green shoots, and listening to the frogs.

***

These incidents happened over 10 years ago. I've experienced many more extreme wet and dry seasons since then. As I write this account, parts of New South Wales (where I work as a conservation biologist) are flooding, and I cannot access my field sites. If I was to offer advice, it would be to remember that field work rarely runs as planned. Safety of yourself and your field assistants is paramount, followed by the safety of study animals. Data are also important, but are the last consideration in that trio. My PhD supervisor once said to me, 'Protect people, protect the animals and protect the data – in that order'. Also, if you are a stubborn individual like me who doesn't like to ask for help, when you're facing tricky decisions and situations try to remember that other people are generally more willing to help than you realise. Sometimes, you just need to (1) ask for their help; (2) be willing to acknowledge your mistakes or lack of knowledge; (3) be prepared to laugh at yourself. Also, ALWAYS stick to the track!

Acknowledgements

First of all, thank you to my partner Matt Gill who has supported me (and often accompanied me) through all these years of wildlife research. Thank you also to Dr S. Topa Petit who expertly supervised my PhD and provided helpful edits on an early version of this chapter.

INDEX TO LOCATIONS AND SPECIES